AF248661

Uses of Enzymes
and
Immobilized
Enzymes

Uses of Enzymes
and
Immobilized
Enzymes

Francis X. Hasselberger

Nelson-Hall, Chicago

Library of Congress Cataloging in Publication Data

Hasselberg, Francis X.
Uses of enzymes and immobilized enzymes.

Bibliography: p.
Includes index.
1. Enzymes. 2. Immobilized enzymes.
3. Enzymes—Therapeutic use. 4. Enzymes—
Industrial applications. I. Title [DNLM:
1. Enzymes. 2. Enzymes—Therapeutic use.
3. Enzymes, Immobilized   QU135.3 H355u]
QP601.H3224   574.1'925   78-8498
ISBN 0-88229-345-1

Manufactured in the United States of America.

10   9   8   7   6   5   4   3   2   1

Contents

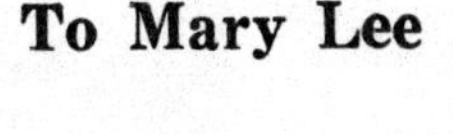

To Mary Lee

Preface

Enzymes are among the most fundamental materials which make life possible; yet, the average person has little familiarity with these biological catalysts. Few books have appeared which are intended to inform the public on this topic. With the rather recent development of immobilized enzymes, the increased use of these active proteins is assured. The intelligent, nontechnically trained person has a need to know of this important and interesting area of research. A large chemical company used to advertise, "Better things for better living—through chemistry." And chemistry has improved the quality of life in the United States. Enzymes have contributed to this improvement and the best is yet to come. Only in recent years has man learned how to handle these molecules properly. The greater knowledge of the fundamental properties of enzymes has made their exploitation by man more beneficial. The recent development of uses of enzymes in cancer therapy is a striking example of this. This same example also answers the question: Why is it important for the intelligent nonscientist to know about this area of research?

Enzymes are fundamental to life. Plants, exposed to sunlight, use carbon dioxide and water to synthesize carbohydrate; nitrogen from the soil is used to synthesize protein. Both of these classes of natural products are synthesized by a series of enzyme-catalyzed reactions. Animals (and man) eat the plants and make meat—also by a series of enzyme-catalyzed reactions. Man eats animal protein. He decomposes it to amino acids by a series of enzyme-catalyzed reactions and then builds human protein in another series of enzyme-catalyzed steps. These natural uses are discussed early in the book, and they lead to a chapter on fermentation (enzyme reactions in which microorganisms are used as a source of enzymes).

Immobilized enzymes are introduced early and their uses are described throughout the book whenever the opportunity arises. This is the first book written for the informed layman dealing with this topic.

Industrial uses are described and it is shown how the enzymic process relates to the rest of the industry. Separate chapters are devoted to the food and dairy industries.

Enzymes are useful in analysis and medical diagnosis. Two chapters describe these uses and form a "bridge" between sections dealing with industrial and medical applications. The second of these chapters introduces a device known as an enzyme electrode—the result of the happy merger of two rapidly growing areas of technology—the immobilized enzyme and the specific-ion electrode.

The medical chapters show the great promise of enzymes in treating a number of particularly horrible diseases. After a rather general chapter showing the effectiveness of enzymes in treating everything from viral diseases to athletic injuries, there are chapters describing uses of enzymes in cancer therapy and the treatment of diseases that arise when enzymes are missing from the human body. It will be difficult to read the chapter on cancer therapy without wondering why American cancer patients die without having the opportunity to benefit from enzyme therapy. In a chapter dealing with the antileukemia enzyme asparaginase, we see an accidental discovery and a planned research program and how one can lead to the other.

The synthesis of enzymes in the laboratory was an epic achievement and it is described in nontechnical terms. The book is concluded with a potpourri of applications and the author's speculation on the future of enzyme uses.

It is the thesis of this book that science can be written in a language understandable to an intelligent, nontechnically trained person, and that, when it is, it is extremely interesting. A rapid reading of this book will provide a good overall background in the uses of enzymes as well as a working knowledge of immobilized enzymes. A careful second reading will probably give a greater depth of appreciation.

The names of enzymes are rather like the names of characters in Russian novels. The reader should not be unduly concerned by this problem. It is more important to understand that a given reaction is catalyzed by an enzyme than to be concerned with the name of the enzyme.

Chemical equations are provided for those who feel comfortable with them. If they are more of a cause for concern than a help, ignore them. The general message of the book can still come through without them.

This book is intended primarily to serve as an introduction to uses of enzymes and immobilized enzymes for the intelligent, nonscientifically trained reader. But it also has a secondary purpose—to supplement courses in biochemistry, biology, and chemistry. There is little time available in these courses for consideration of the practical applications of enzymes discussed in this book. Yet, a knowledge of these applications is certainly desirable for science students. An awareness of practical applications may lead to a greater appreciation of theoretical concepts and lead the student to devise additional applications as his knowledge expands. The book is therefore suggested as an addition to the student's required reading list.

The author wishes to express his gratitude to those scientists who kindly provided reprints of their publications and to those corporations which provided industrial literature. Thanks are due to Dr. Samuel Bessman of the University of Southern California, Dr. Roscoe Brady of the National Institutes of Health, Dr. Thomas Chang of McGill University, Dr. Donald Cole of Port Washington, New York, Dr. Louis Goodson of Midwest Research Corporation, Mrs. Mary Lee Hasselberger of the Nebraska State Agriculture Laboratory, Dr. Samuel Huang of the University of Connecticut, Dr. Henry Lardy of the University of Wisconsin, Dr. Gerald Murphy of Roswell Park Memorial Institute, Dr. Sam Pennington of East Carolina University, Dr. Joseph Roberts of Sloan-Kettering Institute for Cancer Research, Dr. George Sensabaugh of the University of California at Berkeley, Dr. Paul Wolf of the University of California at San Diego, the late Dr. Max Wolf of Miami Beach, and Dr. Oskar Zaborsky of the National Science Foundation. Industrial literature, patents, or other publications were provided by American Cyanamid Co., Corning Glass Works, Dairyland Food Laboratories, Fermentation Design, Inc., General Foods Technical Center, Inolex Corporation, Leeds and Northrup, Miles Laboratories, Monsanto Co., Pfizer, Rohm and Haas, United States Department of Agriculture, Northeastern Regional Research Laboratory, Wallerstein Division of Travenol Laboratories, and Wilson Laboratories. The Mucos Co. of West Germany has been particularly helpful. Drs. Karl Ransberger, E. Heuer, and Wolfgang Werk have provided abundant literature and answered questions in a most cooperative and helpful way.

A word of appreciation is also due to Mrs. Nancy Morton of St. Charles, Missouri, and to Sergeant Donald J. Anderson of Fort Carson, Colorado, for critical readings of portions of the manuscript and to my wife, Mary Lee (Kroner) Hasselberger for reading the entire manuscript and making many helpful suggestions.

Finally, I wish to thank Mr. Paul D. Evans of Lincoln, Nebraska, who provided copies of the manuscript.

Chapter One

An Introduction to Enzymes

THROUGHOUT THIS BOOK, we discuss practical applications of enzymes and immobilized enzymes. After this introductory chapter, we briefly consider a few of the ways in which nature uses enzymes, but then we turn to a consideration of man's uses of these natural wonders. That is the real subject matter of this book.

It is not necessary to have any particular scientific background to understand the uses of enzymes. But, because a basic idea of what enzymes are and how they work will provide a greater appreciation of their practical applications, in this chapter we give a brief introduction to the science of enzymology. Chemical equations are included for those who do have more scientific background; however, they are not necessary for an understanding of the general message of the book.

DEFINITIONS

We introduce enzymes by defining them as proteins which catalyze chemical reactions. A catalyst is a substance which promotes chemical change without itself being consumed in the reaction. It is used in very small amounts (called catalytic amounts) compared with the amounts of reactants which are consumed in the reaction. A protein is a very large molecule made up by bonding together many much smaller molecules called amino acids.

1

Figure 1–1 Amino acids commonly found in enzymes.

Glycine
H_2NCH_2COOH

Alanine
$H_2NCHCOOH$
CH_3

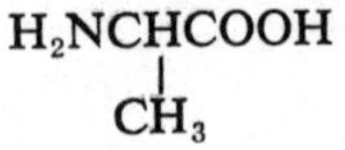

Valine*
$H_2NCHCOOH$
CH
H_3C CH_3

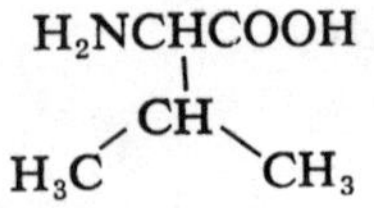

Leucine*
$H_2NCHCOOH$
CH_2
CH
H_3C CH_3

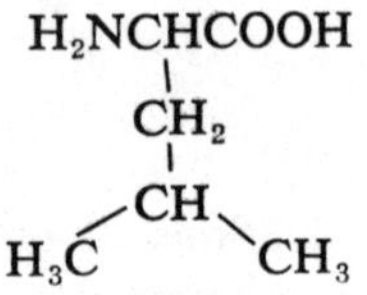

Isoleucine*
$H_2NCHCOOH$
$CHCH_3$
CH_2
CH_3

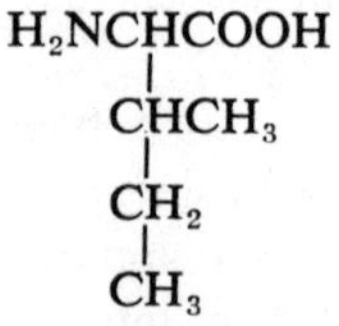

Serine
$H_2NCHCOOH$
CH_2OH

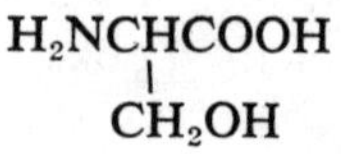

Threonine*
$H_2NCHCOOH$
$HOCH$
CH_3

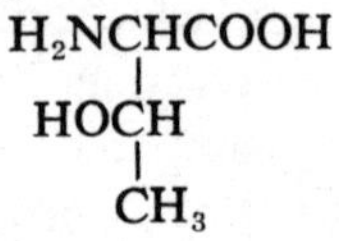

Aspartic acid
$H_2NCHCOOH$
CH_2
$COOH$

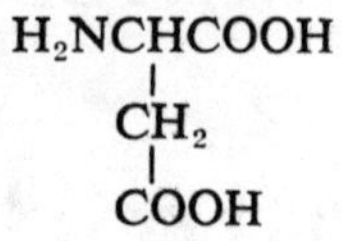

Asparagine
$H_2NCHCOOH$
CH_2
$CONH_2$

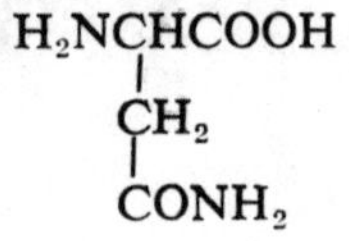

Glutamic acid
$H_2NCHCOOH$
CH_2
CH_2
$COOH$

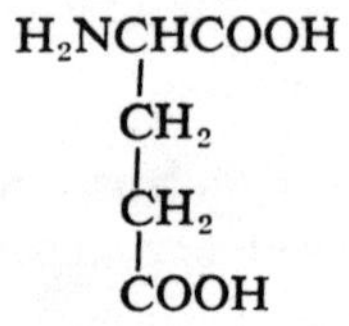

Glutamine
$H_2NCHCOOH$
CH_2
CH_2
$CONH_2$

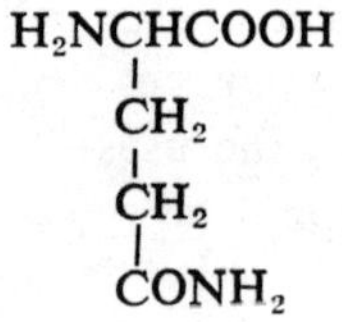

Lysine*
$H_2NCHCOOH$
$CH_2CH_2CH_2CH_2NH_2$

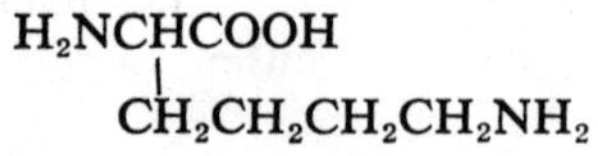

Histidine*
$H_2NCHCOOH$
CH_2
HN N

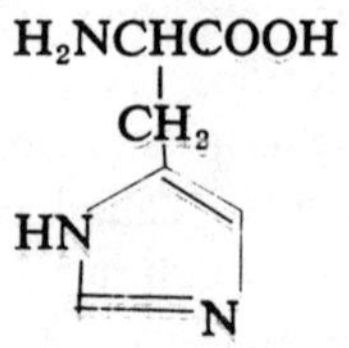

Arginine*

$$H_2NCHCOOH$$
$$(CH_2)_3NHCNH_2$$
$$\|$$
$$NH$$

Cysteine

$$H_2NCHCOOH$$
$$CH_2SH$$

Phenylalanine*

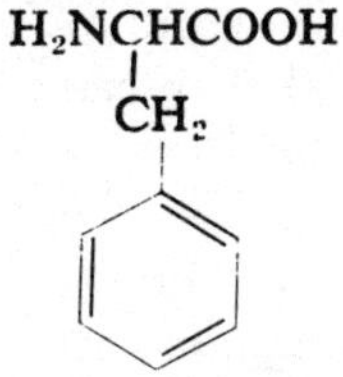

$$H_2NCHCOOH$$
$$CH_2$$

Cystine

$$H_2NCHCOOH$$
$$CH_2$$
$$S$$
$$S$$
$$CH_2$$
$$H_2NCHCOOH$$

Tyrosine

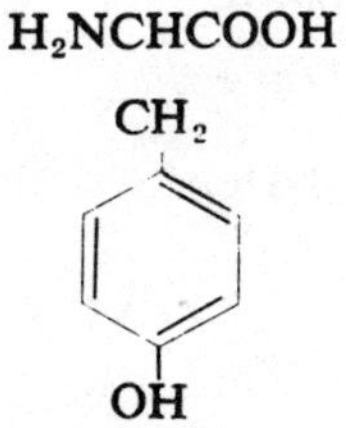

$$H_2NCHCOOH$$
$$CH_2$$
$$OH$$

Methionine*

$$H_2NCHCOOH$$
$$CH_2CH_2SCH_3$$

Tryptophan*

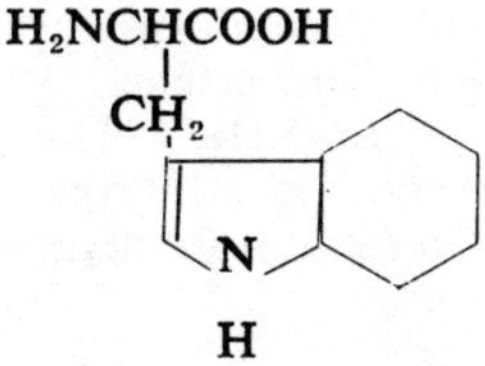

$$H_2NCHCOOH$$
$$CH_2$$
$$N$$
$$H$$

Proline

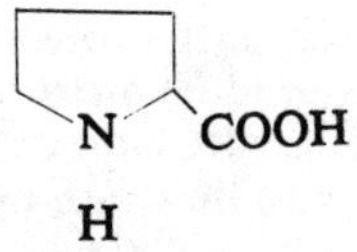

$$N \quad COOH$$
$$H$$

*Denotes amino acids which are essential in nutrition.

COMPOSITION OF PROTEINS

There are 21 of these amino acids which occur commonly in enzymes, and the chemical structures of these molecules are shown in Figure 1–1. Notice that the amino group ($-NH_2$) is attached to the carbon atom adjacent to the carboxyl group ($-COOH$) in all cases. Amino acids having this makeup are called alpha amino acids.

It should be noted in passing that our nutritional need for protein is really a need for amino acids, which are normally obtained by ingesting proteins. A number of amino acids are essential in the diet—that is, they are needed for building body proteins and they cannot be synthesized by the body from other available materials. The essential amino acids are indicated in Figure 1–1.

Proteins are built up by combining the amino group of one amino acid with the carboxyl group of another, eliminating a molecule of water in the process. This is illustrated in Equation 1–1.

Equation 1–1   Formation of the peptide bond.

$$H_2NCH_2COOH \text{ and } H_2N\underset{\underset{\displaystyle CH_3}{|}}{C}HCOOH \rightarrow H_2NCH_2CONH\underset{\underset{\displaystyle CH_3}{|}}{C}HCOOH \text{ and } H_2O$$

Glycine and Alanine $\rightarrow$ Glycylalanine and Water

Using the R and R′ notation this becomes:
$$RCOOH \text{ and } R'NH_2 \rightarrow RCONHR' \text{ and } H_2O$$

In the absence of protecting groups, all possible products would be formed. Thus, the product mixture would also contain alanylglycine (R′CONHR), glycylglycine (RCONHR), and alanylalanine (R′CONHR′). This equation is shown only to illustrate the formation of the peptide bond. Protecting groups are discussed in Chapter 14. These groups direct the formation of a single product.

(Just as mathematicians use X and Y to indicate anything they want it to, chemists often use R in this manner.) In Equation 1–1, and throughout the book, R and R′ are used in this manner to indicate the parts of the molecule which are not directly involved in the reaction. In splitting out water from $-NH_2$ and $-COOH$ groups, a bond called a peptide bond ($-CONH-$) is formed. Proteins are composed of very many amino acids all held together by peptide bonds. Ribonuclease, the first enzyme to be synthesized in the laboratory, contains 124 amino-acid residues (amino acids with a water molecule split out in forming the bond). This is a very small protein. Some protein molecules contain several hundred amino-acid residues.

Chemists have measured the atomic weights of the chemical elements, and these weights are defined relative to oxygen having an atomic weight of 16 or carbon having an atomic weight of 12. The sum of the atomic weights of all the atoms in a molecule is the molecular weight. Ribonuclease has a molecular weight around 14,000. Many protein molecular weights are in the hundreds of thousands.

It should be noted that in all amino acids except glycine, there are four different groups attached to the carbon atom that carries the amino and carboxyl groups. In addition to these groups, there is a hydrogen atom and an "R" group. There are two possible nonequivalent orientations of these four groups and the two structures are related as mirror images. Molecules having one of these orientations are called D–forms; those having the opposite orientation are L–forms. The two forms are called "optical isomers." They can be distinguished from one another by the fact that they rotate plane-polarized light in opposite directions. This is detected by use of an optical instrument called a polarimeter. When molecules having these two possible orientations are synthesized chemically, equal amounts of the D– and L–forms are made. Such a mixture is called a "racemic mixture." Obtaining pure D– or pure L–compounds from a racemic mixture is a very tedious job. Processes involving enzymes quite often can be used to obtain a pure L–compound or a pure D–compound. All of the amino acids in proteins are of the L–form.

SOME PROPERTIES OF ENZYMES

We have said that enzymes are proteins that catalyze chemical reactions. One of the most important properties of these catalysts is their specificity. That is—their catalytic effect is limited to a single kind of molecule, or to a relatively small group of related molecules. The molecules which are affected by enzymes are called substrates. Quite often, an optical isomer of the L–form will be a substrate for an enzyme, while the D–form is not. The enzyme is then said to be specific for the L–form.

Proteins are composed of amino acids joined together in long chains by peptide bonds. Sometimes there are two or more of these long chains in an enzyme molecule. The chains are usually held together by disulfide (–S–S–) bonds. The amino acid chains are referred to as the "primary structure" of the protein. There are also interactions between various groups in the protein molecule—groups that are not chemically bonded to each other. Such interactions are referred to as the secondary and tertiary structure of proteins. Additional interaction between complete protein molecules is called quaternary structure. The interactions which make up these higher-order structures are easily disturbed by influences which do not affect the primary structure. Such influencing factors may be heat, organic solvents, or high acidity or alkalinity. Such influences are said to "denature" the protein. Thus, instability is a property of most proteins. A familiar example of protein denaturation is seen in frying an egg. The colorless "egg white" actually turns white when it is heated. This, as well as the simultaneous hardening of the egg, is the result of protein denaturation.

Classification and Naming of Enzymes

All known enzymes have been classified into six fundamental groups based on the types of reactions they catalyze. These groups are: oxidoreductases, transferases, hydrolases, lyases, isomerases, and ligases. An oxidoreductases catalyzes an oxidation or reduction. Oxidation may be thought of as adding one oxygen or removing two hydrogen atoms from a molecule; reduction is the opposite process. No molecule can be oxidized without the simultaneous reduction of another molecule. A transferase catalyzes the transfer of a chemical group from one molecule to another. Hydrolases are probably the most frequently discussed enzymes in this book. Hydrolytic reactions catalyzed by these enzymes are ones in which chemical bonds are broken with the addition of water—hydrogen (H) adding to one of the resulting fragment molecules and hydroxide (OH) adding to the other. This is the reversal of the process we saw in peptide bond formation, where water was split out when the bond was formed. Lyases catalyze the cleavage of chemical bonds without adding water to the resulting fragments. Isomerases catalyze the rearrangement of molecular structures converting a molecule into a related molecule, which is said to be an isomer of the original. A conversion of a D–form to an L–form would be an isomerization. Finally, a ligase catalyzes the formation of a chemical bond between two molecules.

There is a systematic method for naming enzymes that is based on these six fundamental categories. Quite frequently, in scientific papers, authors will mention the systematic name once and thereafter use a more familiar "trivial" name. We shall have occasion to discuss an enzyme systematically named beta–D–galactoside galactohydrolase, but we shall use its trivial name, lactase, or a somewhat more elegant name, beta-galactosidase.

Many enzymes are named by adding the suffix -ase to the name of the substrate. We have already mentioned lactase—an enzyme which hydrolyzes milk sugar—lactose. There are frequently used trivial names which bear no relationship to either the substrate or the reaction catalyzed. These include papain, trypsin, lysozyme, and many more that we shall see in this book.

But the enzyme names should not be allowed to become stumbling blocks; the important idea is that an enzyme catalyzes these reactions. As far as names go, a rose is a rose.

FACTORS AFFECTING ENZYME ACTIVITY

Catalysts are used to accelerate the rates of chemical reactions. The catalytic effectiveness of enzymes depends very much upon the environment in which the enzyme acts. Three important determinants are the temperature, the concentration of substrate, and the acidity or alkalinity of the solution.

In general, the rate of a chemical reaction is approximately doubled by increasing the temperature by 10°C. This applies to enzyme-catalyzed reactions with the limitation that excessively high temperatures denature the protein, destroying the activity. Thus, the rates of enzyme reactions increase with increasing temperature, until a point is reached at which the protein becomes unstable. Then, the activity decreases with increasing temperature due to enzyme denaturation. The temperature at which the enzyme-catalyzed reaction goes most rapidly is called the optimum temperature for that enzyme.

The rate of reaction depends on the substrate concentration. Since the enzyme is used in catalytic amounts, the substrate concentration is much greater than that of the enzyme. By increasing the concentration of substrate, the reaction rate increases, until a point is reached at which the enzyme is "saturated" with substrate—it can handle no more in a given time. At that point, the addition of more substrate does not accelerate the reaction. The rate becomes independent of substrate concentration.

Just as high temperatures can denature enzymes, so can excessive acidity or alkalinity. These factors also affect enzyme reaction rates. Chemists measure acidity and alkalinity on a scale known as pH. For our purposes, it can be considered a 15-step scale ranging from pH O to pH 14. A pH of 7 is neither acidic nor alkaline; it is said to be neutral. Pure water has a pH of 7. An acidic solution has a pH below 7; the farther below 7 it is, the more acidic it is. An alkaline solution has a pH greater than 7; the higher the pH, the more alkaline the solution. The human mouth has a pH near neutrality—between 6 and 8. The human stomach uses hydrochloric acid as a digestive juice. It has a pH around 1. Each enzyme responds characteristically to pH. Some are most active in neutral solution, whereas some need acidic conditions and others alkaline. The range of pH over which an enzyme is active may be very broad or very narrow, but each enzyme has an "optimum pH." That is, the pH at which the enzyme-catalyzed reaction occurs most rapidly.

Additional factors in determining enzyme reaction rates are cofactors, metal ions, and inhibitors. The first two of these generally increase enzymic activity, although some metal ions fall into the category of inhibitors—materials which retard or terminate enzyme activity.

Cofactors (or coenzymes) are relatively small organic compounds which are required for the activity of many enzymes. Many cofactors are converted by the enzyme reaction to a form which no longer has the power of assisting the enzyme in catalyzing the reaction. Therefore, the reaction mixture must contain one molecule of cofactor for every molecule of substrate to be converted. Many of the vitamins which are essential to human health function as part of enzyme cofactors.

Many enzymes require the presence of metal ions for activity. A metal ion is an atom of metal which has lost one or more electrons and therefore has a positive electrical charge. Table salt (sodium chloride) contains sodium ion, for instance. The metal ions most frequently used as enzyme activators are magnesium, manganese, zinc, potassium, iron, copper, calcium, cobalt, and molybdenum.

Inhibitors are atoms, ions, or molecules which retard or terminate enzymic activity. They are classified as competitive and noncompetitive. A competitive inhibitor is a substance which combines with the active site of the enzyme, preventing the substrate from having access to that site and thus preventing reaction. The active site of the enzyme is the part of the enzyme molecule which must combine with the substrate before any conversion is affected. A competitive inhibitor blocks this site.

A noncompetitive inhibitor combines with the enzyme at a location other than the active site. It does not affect the bonding of the substrate to the enzyme, but nevertheless it retards the conversion of the substrate to product.

Enzyme reactions in general depend upon an aqueous environment. Organic solvents can inhibit enzymes and, in sufficient concentrations, they precipitate the proteins.

Certain metal ions, especially ions of heavy metals such as mercury, lead, and barium, inhibit enzyme reactions. The sulfhydryl group (–SH) is part of the active site of many enzymes and the heavy metals frequently combine with this reactive group.

Certain enzyme inhibitors have been used as insecticides and have been considered for use as war gases. These are the organophosphorus compounds. They are discussed in Chapter 15.

SOURCES OF ENZYMES

All living organisms depend on enzymes, so all living organisms may be considered sources of enzymes. This includes microorganisms, plants, animals, and man. In this book, we shall see examples of uses of enzymes obtained from all of these sources.

Microorganisms are an excellent source of enzymes not only because they can be grown in small quantities or in industrial scale quantities, but also because many of these enzymes are secreted by the microorganism into the medium in which they grow. This greatly facilitates the isolation of the enzyme.

Plant enzymes are usually obtained by grinding the plant material and extracting the ground vegetable material to separate the soluble protein from the insoluble matter. Animal enzymes are similarly obtained. Usually, it is determined which organ is particularly rich in the desired enzyme and that particular organ is used as the source of the enzyme.

We shall see examples of uses of human enzymes derived from placental tissue and from the urine. Both of these are considered in the medical section.

PURIFICATION OF ENZYMES

It can be seen that the enzymes are derived from complex materials and that they will be mixed with many types of soluble materials. This necessitates complex purification procedures. The degree of purification will depend, to a large extent, on the intended use of the enzyme. An enzyme used to treat cowhide in leather manufacturing does not need the degree of purification required by a medicinal enzyme preparation, for example.

Enzyme purifications are often based on selective precipitations using organic solvents (mostly alcohols and acetone) in various concentrations and inorganic salts (most frequently ammonium sulfate) in progressively increasing concentrations. A relatively new physical technique known as gel permeation chromatography (or gel filtration) has been very useful in enzyme purification. Affinity chromatography (discussed in Chapter 15) exploits the specificity of enzymes to affect their purification. An electrical technique called electrophoresis separates proteins on the basis of their electrical charges. This technique can even separate enzymes which catalyze the same reaction, yet have different chemical compositions, and, consequently, different electrical charges. Such enzyme forms are called isoenzymes. Electrophoresis and isoenzymes are discussed in Chapter 8.

The development of purification techniques has been largely responsible for the tremendous gains in knowledge in the enzyme field in the last 30 years. Biochemists, studying enzymes in dilute aqueous solution, have had progressively purer catalysts for their studies. This greater purification has also led to more practical uses. More enzymes are being made available pure enough for use in treating sick humans and animals and, as we shall see, the results are quite impressive.

ANTIGEN-ANTIBODY REACTIONS

We shall have occasion to mention the antigen-antibody reaction in the medical section. It is another example of a specific interaction between a protein molecule and another molecule. The protein is an antibody, not an enzyme. The second molecule is an antigen, not a substrate. Frequently, the antigen is a "foreign" protein introduced into the body. It is recognized as foreign and the body reacts to remove it. To do this, proteins known as globulins form antibodies—proteins which react specifically with the foreign substance which caused their formation. Usually, an insoluble complex is formed and is easily removed from the body.

This should be enough background for a good understanding of the rest of the book. We shall now briefly mention some natural uses of enzymes before turning to our principal topic—man's uses of enzymes.

Chapter Two

Some Natural Uses of Enzymes

In considering "natural uses of enzymes," it is fitting to mention that all forms of life—man, animals, and plants—are completely dependent upon enzymes. Life, as we know it, is not possible without these bioactive proteins. All forms of life exist as cells. Many microorganisms consist of a single cell; a human being contains billions of cells. All living cells have the common feature of metabolism—a feature mediated by enzymes.

Plant cells differ from animal cells in that they have a cell wall, commonly composed of cellulose; the animal cell has a cell membrane, but no rigid wall. Yet, the basic components of the cell are fundamentally the same, whether the cell comes from a plant or an animal. The common material, called protoplasm, is a jellylike, viscous substance, containing 70% to 90% water, 10% to 25% organic matter, and 1% inorganic ash.

The total amount of protoplasm that has ever lived on earth is many times greater than the total mass of the planet. Since the law of conservation of mass states that matter is neither created nor destroyed, it becomes apparent that matter must be reused over and over again to provide the components of living creatures. A consideration of two cycles of nature will serve to exemplify the means by which this reuse is achieved. We shall consider the carbon cycle and the nitrogen cycle.

THE CARBON CYCLE

By the process of photosynthesis, green plants take many tons of carbon dioxide out of the air each day. Yet, the total concentration of carbon dioxide in the air remains essentially constant at approximately 0.035%. Living plants and animals return small amounts of carbon dioxide to the atmosphere by respiration, but quantities are meager compared with the quantity removed by photosynthesis. The vast majority of carbon dioxide regeneration is due to the activity of decay microorganisms, which decompose the carbon compounds of dead plants and animals, converting them into carbon dioxide and water. The plant takes carbon dioxide from the air and uses it to build plant tissue. The plant is consumed by animals or man and the tissue is converted to animal or human tissue. The death and subsequent decomposition of plants, animals, and man returns carbon dioxide to the air, completing the cycle.

THE NITROGEN CYCLE

Every farmer knows the importance of nitrogen in growing his crops. The nitrogen in his fertilizer is used for the synthesis of amino acids and proteins in the plant. Plant protein is consumed and converted into animal protein. Excess protein is decomposed and excreted as urea, uric acid, or ammonia. Decay bacteria convert urea and uric acid to ammonia. A combination of nitrite and nitrate bacteria convert ammonia to nitrate, which again serves as a fertilizer for the soil. Some ammonia, nitrite, and nitrate are converted to nitrogen by denitrifying bacteria.

Atmospheric nitrogen can be "fixed" (converted to organic nitrogen compounds) by certain soil bacteria and by combinations of legumes (such as peas and soybeans) with other bacteria. Ultimately, the plant materials and the nitrogen-fixing bacteria fall prey to decay bacteria. The nitrogen-containing components are converted to ammonia, thence to nitrate, which reenters the nitrogen cycle.

It would be easy to describe similar cycles for other important elements, such as oxygen and phosphorus, but the general pattern is the same. An element is taken from some nonliving source, used by living plants and animals, and returned to the source after the cessation of life. "Remember, man, that thou art dust, and unto dust thou shalt return."

It may seem as though we have digressed from our consideration

of enzymes, but this is more apparent than real. None of the processes
we have mentioned takes place, except by enzyme action.

PHOTOSYNTHESIS

Photosynthesis is the process by which green plants (and some
purple ones) are able to synthesize organic compounds from carbon dioxide and water. The presence of chlorophyll (or other pigments) in
these plants gives them the power to absorb sunlight. It is this use of
solar energy which makes possible the conversion of an estimated 300
million tons of carbon dioxide to carbohydrate each year. Yet, not one
gram of this is converted without enzymes.

It was established in the eighteenth century that photosynthesis
occurs only in the green part of plants and that light is essential for the
process. The green color is confined to small subcellular organelles
(particles) called chloroplasts. The electron microscope has revealed
that there are even smaller particles, called grana, inside the chloroplasts. These grana contain the chlorophyll and they are the sites of
photosynthesis.

The rates of ordinary chemical reactions are approximately
doubled for every 10-degree rise in temperature, but photochemical
(light-catalyzed) reactions are essentially independent of temperature.
Early in the twentieth century, researchers noticed that an increase in
temperature in bright light leads to an increased rate of photosynthesis.
Yet, if light intensity is low, temperature changes do not have much effect on the rate of photosynthesis.

It was concluded that photosynthesis is not a strictly photochemical process and that there must be a purely chemical (dark) reaction to
explain the acceleration of rate with increasing temperature in bright
light. The investigators further concluded that the photochemical reaction must be capable of limiting the dark reaction, since temperature
has little effect if light intensity is low. It is now known that the absorption of light and the consequent photochemical reaction produce cofactors essential for the dark phase reactions. If these cofactors are in low
concentration, as they are in dim light, the reaction rate is limited, regardless of temperature.

We now know that there are two processes catalyzed by light in
the photochemical sequence. In one of these, water is split to yield reactive compounds, which chemists call oxidizing and reducing substances. The product formed by the oxidizing substance is oxygen gas,
which is given off by plants in photosynthesis. The reducing substance is

used to make the reduced form of the coenzyme, nicotine adenine dinucleotide phosphate (NADPH). This is a cofactor needed in the enzyme-catalyzed "dark reactions."

The second light-catalyzed reaction is known as photophosphorylation. In this reaction, the energy of sunlight is converted to chemical energy by the formation of a compound known as adenosine triphosphate (ATP). This is an extremely important compound because it is widely used by biological systems for storing energy. The molecule is composed of three essential parts: a basic component known as adenine, a sugar called ribose, and three molecules of phosphoric acid which are linked together in a chain. The bond joining the terminal phosphate to its neighbor is called a high-energy bond. Energy is required for the formation of the bond, and energy is liberated when the bond is broken. Thus, reactions that produce ATP are energy-producing reactions, and those that use ATP (converting it to adenosine diphosphate, ADP) are energy-using reactions. When energy is needed to drive a biological reaction, it is usually obtained by breaking a high-energy bond, such as the one in ATP.

Much of our knowledge of photosynthesis was obtained by studies using radioactive isotopes. The chemical reactions of these atoms are the same as the reactions of nonradioactive isotopes. But radioactivity provides a label, which makes it possible to follow their paths of metabolism. For instance, in the study of the light-catalyzed reactions, water labeled with radioactive oxygen (O*) was used. Products were analyzed to determine where oxygen of water was used in the photosynthetic process. All of the radioactive oxygen was found in the oxygen gas, which is evolved in photosynthesis; none was found in the carbohydrate formed in the process.

$$CO_2 \text{ and } H_2O^* \quad \xrightarrow{\text{Sunlight: Green plants}} \quad \{CH_2O\} \text{ and } O_2{}^*$$

$\{CH_2O\}$ is an abbreviated form designating carbohydrate. On the other hand, when CO_2^* was used in the process, the labeled oxygen was found in carbohydrate, but not in oxygen gas.

$$CO_2{}^* \text{ and } H_2O \quad \xrightarrow{\text{Sunlight: Green plants}} \quad \{CH_2O^*\} \text{ and } O_2$$

During the years from 1946 to 1953, Dr. Melvin Calvin and associates at the University of California at Berkeley used radioactive compounds, such as C*O$_2$ and KH$_n$P*O$_4$, to study the mechanism of photosynthesis. Their studies revealed the "carbon reduction cycle," also known as the "Calvin cycle." Dr. Calvin was awarded the Nobel prize for his work.

The Berkeley scientists used the photosynthesizing single-cell algae known as "Chlorella pyrenoidosa" to study the fixation of carbon dioxide. The algae were studied in a system containing all of the essentials for photosynthesis—a source of illumination, carbon dioxide, inorganic phosphate, and nitrate ions. After a while, radioactive C*O$_2$ was introduced into the system. It became "fixed," that is, it was taken up by the algae and metabolized, just as nonradioactive CO$_2$ was. After a few minutes reaction time, the algae were dumped into boiling methyl alcohol. This denatured the enzymes and stopped all reactions. Products of photosynthesis were extracted from the algae, and they were separated by a technique known as "two-dimensional paper chromatography." The technique not only separates the products, but also serves as one means for identifying their chemical composition. The separated products could be extracted from the paper chromatograms for further analysis. The development of the technique was an extremely important event, since it made it possible to analyze a few micrograms, or less, of dozens of different substances in a single, simple operation.

To determine which of the products contained radioactivity, Calvin used X-ray film in contact with the paper chromatogram. When the film was developed, it had a black spot corresponding exactly in position and shape to spots on the chromatogram. A Geiger-Mueller counter was used to measure the degree of radioactivity of the various products.

Early experiments revealed that compounds, such as succinic and glutamic acids, that rapidly become labeled as a result of CO$_2$* fixation in the dark were not labeled highly in the procedure just described; conversely, several other substances, including 3-phosphoglyceric acid and some sugar phosphates, which do not become labeled to any extent in the dark, were heavily labeled after a 1-minute exposure to C*O$_2$ in bright light.

In order to determine which product became labeled first, it was necessary to stop the reaction after a reaction period of less than 5 seconds. The predominant radioactive product was 3-phosphoglyceric acid. This compound was established as the first stable product of carbon dioxide fixation in photosynthesis.

It was presumed that this three-carbon compound was formed by adding C^*O_2 to a two-carbon compound. No such acceptor could be found, so it became necessary to look for other compounds that might be CO_2 traps.

Further experiments revealed that ribulose diphosphate (RiDP), a five-carbon compound, was the CO_2 trap. The enzyme, ribulose diphosphate carboxylase, catalyzes the addition of CO_2 to RiDP. This is the only reaction by which CO_2 is fixed in photosynthesis. The product of the reaction is an unstable compound containing six carbon atoms and two phosphate groups. This material cannot be isolated because it decomposes spontaneously, yielding two molecules of 3-phosphoglyceric acid.

A number of books have discussed in some detail the research involved in elucidating the mechanism of photosynthesis. Because of the availability of these books, and because of space limitations, we shall abbreviate our discussion of the topic. It is important to stress, however, that all of the reactions of the Calvin cycle—the so-called dark reactions of photosynthesis—are catalyzed by specific enzymes. The photochemical reactions provide NADPH and ATP to drive the dark reactions, but they would avail nothing without enzymes. Figure 2–1 illustrates the Calvin cycle.

Figure 2–1   The Calvin carbon reduction cycle.

Ribulose-1,5-diphosphate and CO_2 → 2-carboxy-3-keto-1,5-diphosphoribitol (unstable)

↑ ↓

Ribulose-5-phosphate — 2 molecules of 3-phosphoglyceric acid

↑ ↓

Ribose-5-phosphate — 3-phosphoglyceraldehyde

+ ↓

Xylulose-5-phosphate — Dihydroxyacetone phosphate

↑ ↓

Dihydroxyacetone phosphate combines with Erythrose-4-phosphate forming Sedoheptulose-1,7-diphosphate — 3-phosphoglyceraldehyde combines with Dihydroxyacetone phosphate to give Fructose-1,6-diphosphate

↑ ↓

Fructose-6-phosphate and Glyceraldehyde phosphate combine to give Erythrose-4-phosphate and Xylulose-5-phosphate ← Fructose-6-phosphate and Inorganic phosphate

Respiration

The word respiration is frequently used in a very narrow sense as a synonym for breathing. To the biochemist, it means much more. It includes the metabolic processes by which an organism uses oxygen to produce energy.

Whether the term, respiration, is used in the narrow sense, or in the broad sense, it is a process mediated by enzymes. The important enzyme in breathing is carbonic anhydrase. In the tissues, where there are high levels of carbon dioxide (CO_2) produced by metabolism, the enzyme catalyzes the reaction:

$$CO_2 \text{ and } H_2O \quad \xrightarrow{\text{Carbonic Anhydrase}} \quad H_2CO_3 \text{ (Carbonic acid)}$$

Carbonic acid is carried to the lungs by the circulatory system. In the lungs, carbonic anhydrase catalyzes the reversal of the reaction it promoted in the tissues:

$$H_2CO_3 \quad \xrightarrow{\text{Carbonic anhydrase}} \quad CO_2 \text{ and } H_2O$$

The carbon dioxide is exhaled after being liberated in the lungs.

If we consider respiration in the broader sense—the mechanism by which an organism uses oxygen to produce energy—we have a multi-enzyme process. The process of respiration is very similar in all forms of life—plants, animals, and man.

It is interesting to compare some of the features of photosynthesis with those of respiration. It is seen that they are somewhat opposite processes. Both occur in subcellular particles: photosynthesis in chloroplasts and respiration in mitochondria. Respiration is used by all living cells to "produce" energy; photosynthesis occurs in chlorophyll-containing organisms and it uses energy (although most of the energy is converted to chemical energy and stored in that form). In respiration, the cells use glucose and oxygen to provide energy; carbon dioxide and water are produced in the process. Photosynthesis uses carbon dioxide and water to form carbohydrate and oxygen. Energy from sunlight is used. Respiration goes on continuously, day and night, and decreases the weight of the organism; photosynthesis occurs only in sunlight and it increases the weight of the organism. Chemically, respiration is a process of oxidation; photosynthesis is a process of reduction—the opposite of oxidation. Both processes depend on enzyme catalysis.

The metabolism of glucose (a six-carbon sugar) is the means by which energy is provided to the cell. The sugar is broken down initially in a process called glycolysis. The steps in the glycolytic sequence are listed in Figure 2–2.

Figure 2–2   The Embden-Meyerhoff-Parnas pathway of glycolysis.

Glucose ⟶ Glucose-6-phosphate ⟶ Fructose-6-phosphate

Fructose-6-phosphate ⟶ Fructose-1,6-diphosphate

Fructose-1,6-diphosphate ⟶ 3-phosphoglyceraldehyde
and Dihydroxyacetone phosphate

Dihydroxyacetone phosphate ⟶ 3-phosphoglyceraldehyde

3-phosphoglyceraldehyde ⟶ 1,3-diphosphoglyceric acid

1,3-diphosphoglyceric acid ⟶ 3-phosphoglyceric acid

3-phosphoglyceric acid ⟶ 2-phosphoglyceric acid

2-phosphoglyceric acid ⟶ Phospho-enol-pyruvic acid

Phospho-enol-pyruvic acid ⟶ Enol-pyruvic acid ⟶ Pyruvic acid

Each step in this sequence is catalyzed by a different specific enzyme. ATP is converted to ADP in steps 1 and 3 of the sequence, and ADP is converted to ATP in steps 7 and 10. If glycogen is the starting material, instead of blood glucose, one ATP is saved, because reaction 1 is not needed. In step 6, NAD is converted to NADH.

Note that, in step 4 of the sequence, the six-carbon sugars are split, giving two three-carbon sugars. Therefore, for each glucose-starting molecule, steps 7 and 10 each give two ATP molecules. Thus, energy is produced in glycolysis.

After pyruvic acid has been formed, the next reactions depend on the conditions in the cell. In fermenting yeasts, pyruvic acid is converted to acetaldehyde, which is, in turn, converted to ethyl alcohol. In muscle, under conditions of low oxygen, pyruvic acid is reduced to lactic acid. A buildup of this metabolite causes fatigue. When an adequate supply of oxygen is available, pyruvic acid is oxidized (instead of being reduced) in the "Krebs tricarboxylic acid cycle," which is illustrated in Figure 2–3.

Each of these 11 steps is catalyzed by a specific enzyme. Several of these enzymes use cofactors, such as NAD or NADP. The cofactors are reduced to NADH or NADPH when substrates are oxidized. (No oxidation occurs without a simultaneous reduction, and no reduction occurs without a simultaneous oxidation.) Oxygen is used to convert these reduced cofactors back to the oxidized state for reuse in the cycle. This process is catalyzed by enzymes containing a molecule called flavin and by proteins called cytochromes. Whether cytochromes are enzymes or not seems to be a matter of semantics. Thirty molecules of ATP (high-energy phosphate) are generated for each glucose molecule metabolized by this respiratory route. It is by this formation of ATP

Figure 2–3   Krebs tricarboxylic acid cycle.

Pyruvic acid is formed by glycolysis. Under normal physiological conditions it exists as the salt pyruvate. The other acids of the cycle also exist in salt form.

Pyruvate and Coenzyme A $\longrightarrow$ Acetyl coenzyme A

Oxaloacetate and Acetyl coenzyme A $\longrightarrow$ Citrate $\longrightarrow$ cis-Aconitate
                                                                              $\downarrow$
Oxaloacetate   Isocitrate
      $\uparrow$ $\downarrow$
Malate   Oxalosuccinate
      $\uparrow$ $\downarrow$
Fumarate $\longleftarrow$ Succinate $\longleftarrow$ Alpha-ketoglutarate

Carbon dioxide (CO_2) is liberated in steps 1, 6, and 7.

that the metabolic process provides energy. Water and carbon dioxide are liberated in specific reactions of the Krebs cycle.

DIGESTION

The respiratory process illustrates one of the important ways of obtaining energy in the cells. Nutrients are broken down to carbon dioxide and water and ATP is produced. The process, by which ingested food is converted to nutrients, which can be metabolized by the cell, is called digestion.

The process begins in the mouth. Food is chewed to break it into smaller particles and to mix it with saliva. The smell, taste, and, frequently, even the thought of food, stimulate the flow of this juice, which serves as a lubricant and provides an enzyme called amylase, which starts the decomposition of starches.

After chewing, the food is transported to the stomach by way of the esophagus, a muscular tube, which automatically undergoes movements known as peristaltic contractions, which push the food down toward the stomach. This organ has two main parts: the highly distensible fundus, which is mainly concerned with the storage of food and the production of digestive juices, and the smaller, more muscular antrum, which churns and mixes the food with the digestive juices. The secretion of gastric juices is stimulated by the autonomic nervous system and by hormones such as gastrin.

The principal digestive substances of the stomach are hydrochloric acid and the proteolytic enzyme, pepsin. The high acidity of the stomach stops the activity of salivary amylase, but makes it possible for pepsin to start the decomposition of proteins. Pepsin does not break proteins down all the way to amino acids, but it breaks the molecules into shorter chains called peptides or polypeptides.

Pepsin is actually secreted as the proenzyme, pepsinogen. This protein is activated by the hydrochloric acid in the stomach, or by previously activated pepsin. The activation seems to involve removing amino acids blocking the active site. Mucus, secreted throughout the gastrointestinal tract, protects the digestive system from being attacked by its own powerful juices.

The stomach is separated from the small intestine by a ring of smooth muscle called the pyloric sphincter. When the stomach contents have been converted to the point where they are mostly fluid, and when the first part of the small intestine is empty, the pyloric sphincter relaxes and the antrum contracts pushing the liquefied food into the small intestine.

The adult small intestine is a muscular tube about 21 feet long. The first 10 to 12 inches are the duodenum. The next 8 feet are the jejunum and the last 12 feet are the ileum.

As the food enters the duodenum, the hormone secretin is released into the blood by the intestinal wall. The blood carries secretin to the pancreas, where it stimulates the flow of pancreatic juices. At the same time, the gall bladder contracts, emptying its contents—bile. The pancreatic juices and the bile are transported to the duodenum by a common duct. Both of these fluids are alkaline and they combine to neutralize the highly acidic fluid emerging from the stomach.

Pancreatic juice is an important source of digestive enzymes. It provides a lipase, which partially splits fats; an amylase, which converts starch to the sugar maltose; maltase, which converts maltose to the simpler sugar, glucose; the proenzymes trypsinogen and chymotrypsinogen; rennin, which clots milk protein; and peptidases, which convert proteins and peptides to amino acids.

The small intestine itself secretes digestive juices containing a number of enzymes. The intestinal juices collectively are called "succus entericus." Among the important enzymes in this mixture are erepsin, which splits peptides to amino acids; amylase, which converts starch to maltose; maltase, lactase, and sucrase, which convert sugars to simpler forms, such as glucose; lipase, which breaks fats down to glycerol and fatty acids; and enterokinase, which activates trypsinogen, converting it to trypsin.

Trypsin activates chymotrypsinogen, converting it to chymotrypsin. Both trypsin and chymotrypsin are proteolytic enzymes, which convert proteins to simpler molecules, but not all the way to amino acids. Aminopeptidases and carboxypeptidases convert proteins and peptides to readily absorbable amino acids.

Bile is important in digestion not only because of its alkalinity, which helps to neutralize the acid coming from the stomach, but also because it contains bile salts. These compounds help to emulsify fats, making them susceptible to attack by pancreatic lipase. They also promote the absorption of fats.

The liver secretes bile at a steady pace. Unless there is chyme (partially digested food) in the duodenum, the duct to the duodenum is kept closed by another circular muscle known as the sphincter of Oddi. This causes the bile to back up and to fill the gall bladder. The presence of fat in the duodenum stimulates the release of the hormone, cholecystokinin, which causes the relaxation of the sphincter of Oddi, as well as the contraction of the gall bladder. Digestion is completed in the small intestine.

The enzymatic processes which comprise digestion convert complex foodstuffs into relatively simple molecules, such as amino acids, glucose, fatty acids, and glycerol. These soluble materials are absorbed into the walls of the intestinal lining. From there, they pass into the extracellular fluid and then into the lymph and plasma. They are delivered to cells for metabolism.

BIOLUMINESCENCE

A somewhat spectacular example of a natural use of enzymes is provided by the phenomenon of bioluminescence—the process by which certain members of the plant and animal kingdoms are able to emit light. The most familiar of this genre is the firefly, but there are luminescent varieties of fishes, sponges, crustaceans, beetles, worms, centipedes, mollusks, sponges, and microorganisms, including bacteria and fungi. Different species emit different colors of light—red, green, yellow, or blue. The railroad worm of Uraguay produces light of two colors; it has a row of green lights along the sides of the body and a pair of red lights at the head.

The production of light by living organisms is mediated by enzymes. The details of the process vary in different species, but, in general, the reaction can be described as the oxidation of a complex organic molecule called luciferin by the enzyme luciferase. Luciferin and luciferase from various species may be quite different from each other chemically, but the reaction always requires oxygen and it is always catalyzed by an enzyme.

Luciferin and luciferase may be extracted from the firefly and mixed in a test tube, along with magnesium ion and ATP. Light will be produced. This is another example of a reaction in which ATP is used as a source of energy.

The Natural State of Enzymes

Much of this book will be concerned with uses of enzymes that have been devised by man. We shall see that enzymes may be obtained from microorganisms, plants, animals, and, occasionally, from human sources. Generally, the enzyme is extracted from the tissue of its origin and purified to an extent that depends on the intended use of the enzyme. For instance, if an enzyme is to be injected into a patient to cure his cancer, it must be purified much more extensively than an enzyme which will be used to remove the hair from cowhide in the leather industry.

Most frequently, the enzymic methods devised by man make use of proteins dissolved in an aqueous (water) solution. This might lead to the impression that enzymes are floating around free in solution in the natural state. Such an impression would be inaccurate.

The majority of enzymes are localized within the cell, usually in subcellular particles, such as mitochondria, chloroplasts, ribosomes, and lysosomes. The cells have a membrane which permits small molecules to diffuse in and out, but which prevents the escape of large molecules, such as proteins. Thus, the enzymes are entrapped within the cell. Frequently, the enzyme is held in a fixed position in the cell. It may be adsorbed to some cell membrane component, or it may be chemically bonded to the cell membrane.

In relatively recent years, scientists have tried to mimic this natural condition of enzymes. This attempt has given rise to the study and use of immobilized enzymes—the subject of the next chapter.

Chapter Three

Immobilized Enzymes

It has been the standard practice of biochemists to study enzyme-catalyzed reactions by using purified enzymes in dilute aqueous solution. Although this practice has served well and has expanded biochemical knowledge immensely in the last 30 years, it represents very unnatural conditions. The environment of the living cell is quite different from a dilute aqueous solution, and the enzyme, far from being in a chemically pure state, is "contaminated" by the presence of other proteins, as well as fatty materials, carbohydrates, and other biological molecules.

Around the middle of the 1950s, a new technique was introduced which makes it possible to study enzymic reactions under conditions more similar to those seen in the living cell. This technique is called "immobilization" of enzymes.

Besides permitting enzyme studies to be conducted under "more natural" conditions, there are a number of practical advantages to immobilized enzymes. Quite frequently, immobilized enzymes are more stable than purified enzymes in dilute aqueous solution. They may be more stable to heat, to broader ranges of pH, and to storage. Proteolytic enzymes, such as trypsin, are stabilized against self-destruction. Trypsin, as a protein, is a substrate which is attacked by proteolytic enzymes, such as trypsin. When the enzyme is immobilized, this self-digestion is essentially eliminated. Proteins are large substrates, and it is generally found that large substrates are attacked slowly, if at all, by immobilized enzymes. The fact that neither the substrate, nor the enzyme, is in solution further reduces the possibility of reaction.

Most frequently, the immobilized enzyme is insoluble in aqueous solution. Thus, it is easily possible to recover the catalyst after it has been used. The same enzyme can be used repeatedly in this manner. It is also possible to use the immobilized enzyme in continuous processes. The immobilized enzyme is packed into a column and substrate solution is pumped through it at a carefully controlled rate and under carefully chosen conditions of pH, temperature, and salt concentration. The process can be run continuously as long as the enzyme remains active. This may well be for a period of several months.

Soluble enzymes are not generally recovered for reuse when batch processes are used; it is a very difficult problem to recover the enzyme for reuse. Except in a few isolated instances, soluble enzymes cannot be used in continuous processes. In these "isolated instances," the enzyme must be restrained by a semipermeable membrane (a membrane which permits small substrate and product molecules to pass, but which prevents escape of large protein molecules). Thus, the enzyme is, in a sense, immobilized.

Methods for Immobilizing Enzymes

There are essentially five basic methods for immobilizing proteins. These are adsorption, entrapment in a polymer lattice, cross-linking, microencapsulation, and chemical bonding. Each method has its advantages and disadvantages. No one method is best for all applications.

Adsorption

By far the simplest method for immobilizing enzymes is physical adsorption. An aqueous solution of enzyme is contacted with some solid material which has adsorptive properties. The solid is then washed to remove any nonadsorbed materials. There are no physical or chemical changes in the adsorbent. The enzyme may change in physical properties and this change may result in a change in chemical properties. For instance, a folded or coiled structure may spread out over the surface of the adsorbent. This physical change may make the enzyme catalytically less active, or even, inactive.

Frequently, adsorption is a reversible process. The bonding of an enzyme to an adsorbent may be quite stable under one set of circumstances, but, under other circumstances, the enzyme may be washed off with ease. We shall see an example of this in Chapter 6. The enzyme aminoacylase, adsorbed on an ionic derivative of cellulose, is firmly bound when dilute solutions of substrate are passed over the material; but more concentrated substrate solutions readily free the enzyme from the support. This can be a disadvantage in that it limits the amount of substrate that can be passed over the enzyme in a given time; on the other hand, when the enzyme loses activity, it can be washed off the support very easily. Fresh enzyme can then be added and the process continued. It is very important to establish optimum conditions for adsorption; this is frequently done by trial and error.

Lattice Entrapment

To understand this method, it is necessary to know a little about polymerization. Many compounds containing chemical double bonds are able to bond to each other forming a linear chain of these molecules, which are called monomers. In this process of joining together, the double bonds are converted to single bonds. Long chains, formed by joining monomers, are called polymers. If the reaction mixture contains only these monomers having one double bond per molecule, only a linear polymer can be formed. These are not useful for enzyme entrapment. If the mixture contains a small percentage (such as 5%) of compounds having two double bonds per molecule, each of these double bonds can become part of different linear chains. Thus, the chains will be chemically bonded together. Such a polymer is referred to as a cross-linked polymer; the compounds having two double bonds are called cross-linking reagents. It is cross-linking that makes enzyme entrapment possible. The structure of the cross-linked polymer has tiny "chemical cages" produced by the cross-linking. The enzyme is entrapped in these cages.

To produce an entrapped enzyme, an aqueous solution of enzyme is mixed with a monomer and a cross-linking reagent. A catalyst is added to start polymerization. After a few minutes, the liquid solution sets to form a relatively solid mass. This is broken into very small pieces and is washed thoroughly to remove any enzyme that is not entrapped. Small substrate and product molecules can pass freely into and out of the tiny chemical cages, but the large enzyme molecule is unable to get out of the cage. The cage, of course, is a three-dimensional structure, which prohibits enzyme escape in all directions. The most frequently used system is the polyacrylamide system, which is illustrated in Figure 3–1.

Figure 3–1   Polyacrylamide entrapment of enzymes.

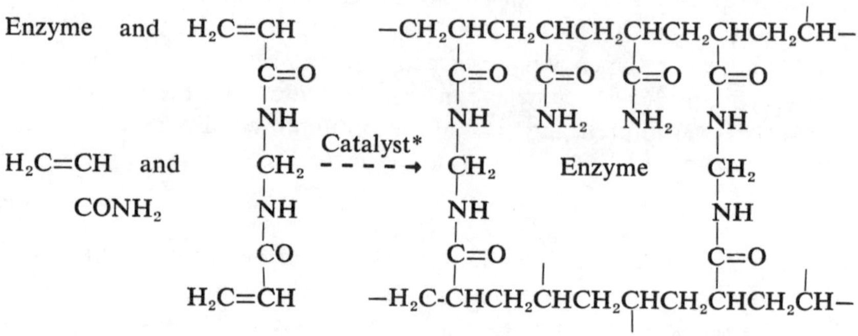

Enzyme and $H_2C{=}CH$ ... $-CH_2CHCH_2CHCH_2CHCH_2CHCH_2CH-$

Acrylamide and N,N′-methylenebisacrylamide
    Catalyst*

-------------- → Cross-linked polyacrylamide
                with enzyme entrapped in matrix

*The catalyst for this reaction may be a chemical reagent, such as persulfate, or an exposure to high radiation, such as X-rays.

The entrapment method is relatively simple, but entrapped enzymes are limited by the fact that only small substrates can be converted by these catalysts. Just as the large enzyme cannot get out of the cage, so large substrate molecules cannot get in.

Cross-linking

If a linear polymer has reactive groups available for chemical bonding, it is possible to treat the linear polymer with a cross-linking reagent and get a cross-linked polymer. The polymers known as enzymes have abundant reactive groups available and these can be used to cross-link the enzyme to another molecule of enzyme, to some inert protein, or to some insoluble carrier molecule. Frequently, enzymes are cross-linked with bovine serum albumin, an inert protein. The most frequently used cross-linking agent is glutaraldehyde. This process produces large, insoluble molecules having enzymic activity. Each protein molecule has many available amino (NH_2) groups. Any number of these can become involved in the cross-linking reaction, depending on the concentrations of polymers and of cross-linking agents. Figure 3–2 illustrates the cross-linking process.

Figure 3–2   Simplified illustration of enzyme cross-linked with glutaraldehyde.

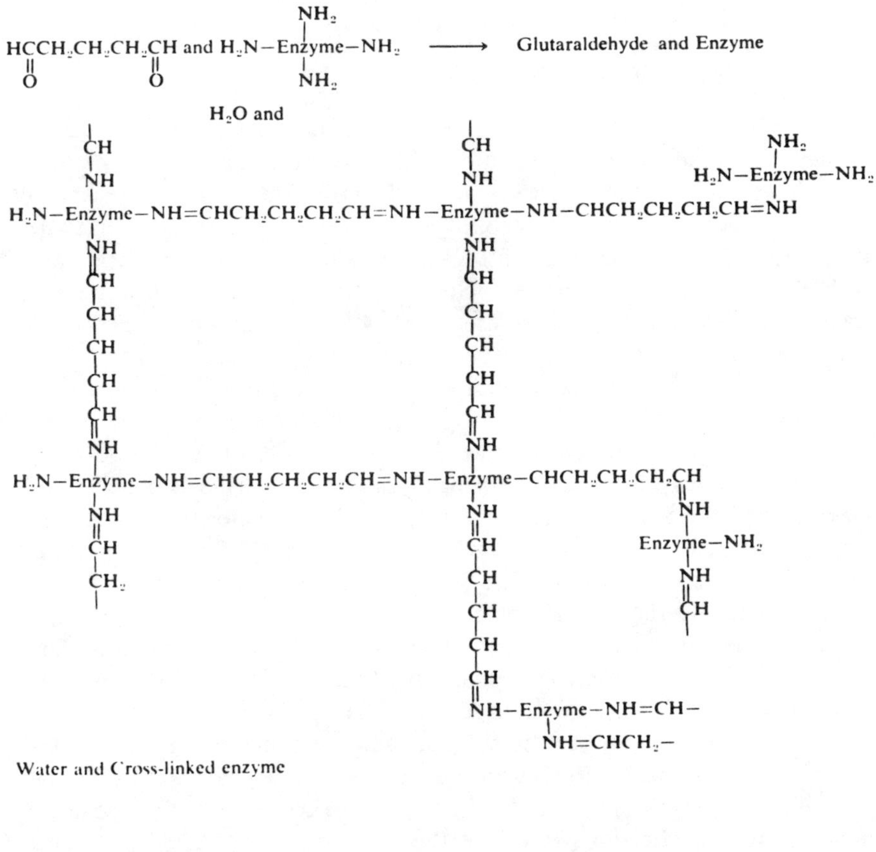

$$HCCH_2CH_2CH_2CH \text{ and } H_2N-Enzyme-NH_2 \longrightarrow Glutaraldehyde \text{ and } Enzyme$$

with the glutaraldehyde carbonyl (O) groups and the enzyme amino (NH_2) groups reacting to give H_2O and the cross-linked enzyme network shown below:

$$H_2N-Enzyme-NH=CHCH_2CH_2CH_2CH=NH-Enzyme-NH-CHCH_2CH_2CH_2CH=NH$$

$$H_2N-Enzyme-NH=CHCH_2CH_2CH_2CH=NH-Enzyme-CHCH_2CH_2CH_2CH$$

$$NH-Enzyme-NH=CH-$$

$$NH=CHCH_2-$$

Water and Cross-linked enzyme

The cross-linking method can be carried out without added carriers. An aqueous solution of enzyme and a cross-linking agent are the only requirements. Indeed, it is not even necessary to have the enzyme in solution. Quiocho and co-workers have cross-linked and studied enzyme crystals. The mechanical strength of the crystals was increased and enzymic activity was retained.

When no support is used, large quantities of enzyme are needed and the cross-linked product tends to exist as a mass of jellylike material. This severely limits the practical use of the method.

An ingenious method, invented by Haynes and Walsh at the University of Washington, combines adsorption and cross-linking. Thus, the enzyme is adsorbed on a support and then is held in place chemically by adding a cross-linking agent. The method has been used with supports ranging from silica to stainless steel.

Microencapsulation

This method is somewhat similar to the lattice entrapment method. It depends upon trapping a solution of enzyme in a molecular cage having pores which readily pass small substrate and product molecules, while preventing enzyme escape. The size limitation is less severe with this method than it is with entrapment. A microcapsule consists of a spherical, ultrathin membrane enclosing a concentrated solution or suspension of enzyme. The size of the pores can be controlled. In lattice entrapment, solid particles are formed and there is no control of pore size.

A typical microencapsulation process is the interfacial polymerization of a molecule containing two amino (NH_2) groups and a second molecule containing two acid chloride groups (COC1). The product of such a reaction is a nylon. The process is called interfacial because the enzyme and the amino compound are dissolved in water, while the acid chloride is dissolved in a solvent which does not mix with water, such as chloroform and cyclohexane. The polymerization occurs at the interface of the water and the second solvent.

A leading proponent of microencapsulated enzymes is Dr. Thomas Chang of McGill University in Montreal. Some of his applications of these catalysts are considered in the medical section.

As Dr. Chang has stressed, a small volume of microcapsules presents an extremely large surface area for contact with substrate solutions. The enzyme may be attached to a carrier, prior to microencapsulation, to increase the stability of the catalyst.

The principal disadvantage of the method is still the size limitation on substrates. Most insolubilized enzymes share this limitation.

Chemical Bonding

This is probably the most frequently used method for immobilizing enzymes. The variety of carriers available for use in this method is virtually unlimited. The carrier may be a natural material, such as cellulose, a synthetic polymer, such as ethylene-maleic anhydride copolymer, or an inorganic material, such as glass or stainless steel.

A few carriers, such as ethylene-maleic anhydride copolymer, are able to react with the enzyme directly; more frequently, however, an activation process is required. The activated carrier is able to bind enzyme but, without activation, it is not.

Although enzymes have several reactive groups, which may be used for chemical bonding, the one most frequently used is the amino (NH_2) group. Hydroxyl (OH) groups, sulfhydryl (SH) groups, carboxyl (COOH) groups and aromatic rings are also used.

The process of activating the carrier consists of introducing onto the surface of the carrier, and, frequently, inside the pores of the carrier, chemical species which can react with the reactive groups of the protein. A large number of techniques are available for doing this. The method is so diverse that it seems inappropriate to attempt to describe any "typical" method.

The chemically bonded immobilized enzyme is adaptable to a wide variety of application techniques. It can be used in any kind of reactor system (see below).

The carrier molecule exerts an effect on the properties of the enzyme. For instance, the optimum pH of an immobilized enzyme is frequently different from that of the free enzyme in solution. If the carrier has a positive charge, the optimum pH will be lower; if it has a negative charge, the optimum pH will be higher than that of the free enzyme. This can be used to advantage; by choosing a suitable carrier, the optimum pH of the enzyme can be shifted from an undesirable value to a more useful one. An example of this is seen in the section on enzymes in dentistry.

Probably the chief disadvantage of the chemical bonding method is that it sometimes uses fairly elaborate activation methods. The percentage of enzyme bonded is sometimes low and large molecules react slowly, if at all, with the supported enzymes. It is believed that the large support structure physically prohibits large substrates from contacting the enzyme in a proper orientation for reaction. Chemists call this "steric hindrance."

SOLUBLE AND COLLOIDAL ENZYME SYSTEMS

So far, we have been concerned only with insoluble immobilized enzymes. But the definition of immobilized enzymes does not demand insolubility. Enzymes chemically bonded to carriers, whether soluble or insoluble, are considered immobilized. There have been relatively few published examples of soluble supported enzymes and, apparently, only one example of a colloidal supported enzyme. These facts permit me to indulge myself to the point of describing some of my own work.

My first experience in the field of immobilized enzymes involved preparing insoluble derivatives of the enzyme, asparaginase, for use in leukemia therapy (see Chapter 12). This was reported in the journal, *Cancer Research.*

Subsequently, soluble and colloidal asparaginase derivatives were sought. A technique, which had been found to be particularly excellent in the preparation of insoluble derivatives, was the activation of cellulose by a compound called cyanogen bromide (CNBr) in alkaline solution. This process had been previously described by Swedish scientists. My first preparation of a soluble supported enzyme used a soluble dextran, which is chemically related to cellulose. It was activated by CNBr to give a highly active adduct with asparaginase. (While insoluble immobilized enzymes retain a relatively small percentage of the activity of the free enzymes, soluble derivatives are usually as active as, and sometimes more active than, the free enzymes.)

When an insoluble support is activated by a chemical, such as CNBr, the chemical is washed off before the enzyme is added. Otherwise, the enzyme would be denatured. To remove the CNBr from the soluble support, a technique called dialysis was used. The solution containing the support, the CNBr, and the alkali was poured into a cellophane-like bag, which was then placed in a beaker of water. The relatively large dextran cannot escape from the bag, but small molecules such as CNBr and alkali easily pass out into the water in the beaker. The water was changed frequently to remove all of the small molecules. The enzyme was then added to the activated support in the dialysis bag. After the reaction was believed complete, the solution was lyophilized (freeze-dried) and the powdery dextran-enzyme adduct was recovered. Free enzyme, free support, and supported enzyme were separated from the reaction mixture by a physical technique known as gel permeation chromatography.

Another investigator at the cancer research center where I worked had reported the preparation of an insoluble derivative of acetylcholinesterase, using a support known as polygalacturonic acid (PGA). This support is related to cellulose and to dextran in that it is composed of sugar molecules bonded together in a very long chain; it differs from those compounds in that it also contains carboxylic acid groups (COOH). The sugar molecules can be activated by CNBr and the carboxylic acid groups can be activated by a class of compounds known as carbodiimides.

The PGA was located and a sample was activated by a carbodiimide in acid solution. Because it was late in the day and because I was tired, I mistakenly washed the solid with an alkaline solution and watched it disappear before my eyes. I poured the material into the beaker and left it on the counter overnight. After some thought, it was obvious what had happened. The next morning, I acidified the material in the beaker and the solid reappeared. This was the first known example of a support, which is easily activated, that is soluble under some conditions and insoluble under other conditions. PGA is soluble above a pH of 4.5 and insoluble below this pH.

Subsequently, PGA was activated in acid solution and washed with acid. It was possible to couple the enzyme to the support in solution, decrease the pH to precipitate the PGA-enzyme adduct, and wash the adduct with acid solution. Alternatively, the dialysis procedure described for the soluble dextran could be used. The process of dissolving and precipitating the support and the support-enzyme adduct was applied to a number of enzymes, but it is limited to those adducts which can withstand the changes of conditions without denaturing the enzyme. This limitation was not severe, since the support stabilizes the enzymes to changes in pH.

Another "accident" led to the preparation of the desired colloidal (gel-like) asparaginase derivative. While paging through *The Merck Index*, I noticed a listing of a compound known as alginic acid. Like PGA, alginic acid contains sugar molecules and carboxyl groups; it also can be "dissolved" and precipitated by a change in pH. But unlike PGA, it does not truly dissolve; instead, it forms a colloidal system.

Alginic acid was activated by carbodiimide in acid solution, just as PGA had been. Asparaginase was coupled to the activated support. Very pronounced light scattering was observed, confirming the colloidal nature of the adduct. The colloidal adduct was easily precipitated by acid. The process was shown to be widely applicable by repeating it with several different enzymes.

Both PGA and alginic acid can be activated by CNBr in alkaline solution. After activation, the supports are precipitated and washed with acid solution to remove excess CNBr. The enzyme is then coupled to form a soluble or colloidal derivative. This process is generally applicable, just as the carbodiimide process is.

During my year at Lehigh University, two undergraduate students repeated some of my work with alginic acid as part of their course work. A publication, describing some of their results, was published in *Biotechnology and Bioengineering* after I left the university. This is the only example of a publication describing a colloidal immobilized enzyme.

Immobilized Enzyme Reactors

The immobilization of enzymes permits their use in continuous processes or their repeated use in batch processes. The hardware, in which the enzymic reactions are carried out, is called a reactor, and there are a variety of types of enzyme reactors. We shall briefly consider a few types of these reactors.

Stirred Tank Reactors

The stirred tank reactor is one of the most common types of enzyme reactors. In its simplest form, it consists of a container with an immersed stirring device. This is the cheapest type of reactor and the most flexible. Many enzyme-catalyzed reactions use, or generate, acid or base. This effect must be counteracted or the enzyme will find itself in an environment of unfavorable pH. This correction is easily made in the stirred tank reactor. There are commercially available devices which detect a change in pH and automatically add acid or base to maintain a constant pH. A device such as this can be used with the stirred tank reactor, but not with other types we shall consider. There are enzymes which require oxygen for maintaining activity. Again, this is easily added to the stirred tank; it is not handled as easily in other types of reactors.

British scientists found that the rate of reactions in the stirred tank reactor varies as the rate of stirring varies. Increased stirring speed accelerates the reaction rate. For this reason, the immobilized enzyme catalyst used in this type of reactor must be durable. A support which disintegrates under the strain of rapid agitation cannot be used.

The stirred tank is the reactor used for batch-type processes. The reactants, cofactors, and activators are mixed in a solution of proper pH, agitation is started, enzyme is added, and the reaction is allowed to proceed. After the reaction is complete, the immobilized enzyme is recovered for reuse in a subsequent reaction. The product is isolated from the reaction mixture.

The stirred tank reactor can also be used in continuous processes. Solution is fed into the reactor on one side while being simultaneously removed from the other side. The entrance and exit ports are covered with a fine mesh material to prevent the escape of the immobilized enzyme. The solution from one stirred tank reactor may be fed directly into a second similar reactor. A series of these reactors may be joined in this manner.

Packed Bed Reactor

Another way to use immobilized enzymes is to pack the catalyst in a column and circulate the substrate through it. Such a reactor is called a packed bed. The substrate solution may be pumped in at the top of the column and the product taken off at the bottom, or the reverse procedure may be used. When substrate is added to the top of the column, there is a tendency for the catalyst particles to jam together, plugging the column. This is particularly true of small particles. The world's first commercial process using immobilized enzymes employs a packed bed reactor with the flow going from the top of the column to the bottom.

Fluidized Bed Reactor

This is another type of reactor in which the immobilized enzyme is used in a column. In the fluidized bed reactor, flow is always from the bottom to the top of the column. In this type of reactor, the column is only partially filled with immobilized enzyme. The catalyst rests on a support at the bottom of the column. When substrate is introduced in a flowing stream, the bed of solid particles is transformed into a fluidlike state. The stream of substrate moves fast enough to overcome the force of gravity acting on the particles, but not so fast as to carry them out of the reactor. The movement of the substrate solution causes the particles to separate from each other and to move about in the column. The column takes on the appearance of a gently boiling liquid. This effect is quite efficient in mixing the catalyst with the substrate.

Hollow Fiber Reactors

Hollow fibers are thick-walled tubes about the diameter of a human hair. The tubes are made of a semipermeable material, such as cellulose, which confines large molecules while freely passing small ones.

In the first hollow fiber enzyme reactor, described by Dr. Peter Rony in 1971, the enzyme solution was placed inside a fiber bundle, which was then temporarily sealed. The substrate solution was introduced to the container outside the hollow fibers. Substrate molecules passed through the fibers to be converted by the enzyme on the inside. Products diffused out. More recent types of hollow fiber reactors have the outside wall of the fiber soaked in enzyme solution. Between this wall and the lumen of the tube is a dense, ultrathin, semipermeable membrane. The substrate solution, pumped through the fibers, can cross this membrane to be converted by the enzyme.

It seems likely that soluble supported enzymes would be useful in this type of reactor. This would provide stabilization of the enzyme without reducing its activity. Soluble cofactors would have to be physically confined to the area of the enzyme, and attachment to a large, soluble carrier, such as dextran, would provide a means for effecting this confinement.

Shoestring Reactor

After the publication of the method for bonding enzymes to cellulose by cyanogen bromide activation, stories circulated of people bonding enzymes to their lab coats, so they could tell of the enzymic activity of their lab coats. (Many lab coats are made of cotton, which is rather pure cellulose.) This caused me to speculate about possibly more practical uses. The "shoestring reactor" is one of the possibilities. While this has never been tested, there is no theoretical prohibition.

A length of cotton material would be activated and exposed to an enzyme solution to bind the enzyme to the cotton. This "shoestring" could be used in a column or in a tank. It would be easy to have a series of strings carrying different enzymes for sequential action. A string which had lost its catalytic activity could be replaced very easily.

Cofactors would be attached to these strings for use with soluble enzyme systems. A cofactor, such as nicotine adenine dinucleotide (NAD), is needed in many enzyme reactions. During reaction, it is converted to a chemically reduced form (NADH) which is no longer catalytically useful in the reaction. But other reactions need NADH as a cofactor, and they cannot use NAD. Thus, the strings could be moved back and forth between reactors, depending on whether NAD or NADH is needed. One reactor would "regenerate" the cofactor for the other reactor.

Immobilized Whole Cells

This provides some introduction to the concept of enzyme immobilization. Throughout the rest of the book, we shall have occasion to refer to uses of immobilized enzymes. It is fitting to close this chapter with a brief mention of a relatively new concept—immobilized whole cells.

Microorganisms, such as yeasts, bacteria, and fungi, are abundant sources of enzymes. But the process of extracting, isolating, and purifying the desired enzyme can be quite laborious and can contribute substantially to the cost of the operation. In recent years, scientists have found that it is not necessary to isolate the enzyme; the whole cell can be used as a source of the enzyme. The same techniques used to immobilize enzymes can be employed to immobilize whole cells. A commercial application of immobilized whole cells is described in Chapter 6. But, now, we shall consider an application of the enzymes found in microorganisms which mankind has used for thousands of years—fermentation.

Chapter Four
Fermentation

ANTON VAN LEEUWENHOEK, in 1680, became the first man to see yeast cells under the microscope. It was not until 1835, however, that Charles Cagniard de Latour in France and Teodor Schwann in Germany saw cells multiplying by budding in the deposit formed in beer fermentation vats. The deposit was, of course, yeast cells. The observation of these reproducing cells caused them to suggest that microorganisms might be involved in fermentation.

The hypothesis was rejected because living organisms were not regularly found in other fermentations, such as lactic acid fermentations. The leading chemist of the day, Justus von Liebig, maintained that brewer's yeast was active, not because it was alive, but because it had been in contact with air. He said that it was the dead portion of yeast, which had been alive and was in the process of decomposition, that acted on sugar.

In 1854, a 31-year-old scientist was appointed professor of chemistry and dean of the new School of Science at the University of Lille in the heart of the French vineyard area. His name was Louis Pasteur.

In 1856, a winery owner, who had suffered serious financial losses because of the souring of his wine, sought Pasteur's help. Although trained as a chemist and physicist, Pasteur attacked the problem using a tool of the biologist. He looked at fermenting juices under a microscope and sketched in a notebook what he saw. All of the alcoholic fermentation liquors contained one type of microorganisms, but those that went sour contained an additional microorganism. This second microorganism seemed to act after the alcoholic fermentation was completed. Pasteur suggested that wine should be heated gently after fermentation is completed to kill any microorganisms remaining. The suggestion was accepted; the souring stopped; the winery was saved. Pasteurization was born.

On August 3, 1857, Pasteur announced to the Lille Society, and to the world, that fermentations were caused by living organisms.

The art of fermentation had been used throughout the history of man. In Genesis, Moses tells us of Noah's use of fermented grape juice. Yet, it was not until 1857 that the science of fermentation made its debut.

Another discovery of epic importance was reported by Eduard and Hans Buchner in 1897. These scientists ground yeast cells with sand to break the cells apart. Juice from the broken cells was squeezed through a cloth bag in a filter press. A thick sugar syrup was added as a preservative. But an immediate and vigorous fermentation began. The Buchners found that the yeast juice—with no living cells—could convert sugar into ethyl alcohol and carbon dioxide, just as living yeasts did. In a sense, Pasteur had been wrong; living cells were not necessary for fermentation. The true ferments, or enzymes, were produced by the microorganisms, but they could be separated from them with retention of activity.

Microorganisms Used in Fermentations

As we now know, fermentations may be thought of as enzymic processes, in which living microorganisms are the sources of enzymes. The microorganisms used in fermentations are chlorophyll-free plants, including bacteria, fungi, yeasts, and actinomycetes.

Since these plants do not contain chlorophyll, they cannot photosynthesize. They must exist on other living organisms as parasites, or use the dead remains of plants and animals for nourishment. Those surviving in the latter manner are called saprophytes. Such organisms synthesize enzymes, which diffuse out of the cell into the environment, changing available complex substances into simpler substances, which can be used as food. Digestion is completed outside the cell and nutrients are absorbed.

Bacteria are the "oddballs" in the group of microorganisms that causes fermentations. All bacteria are one-celled organisms and the cell is relatively simple compared to the cells of other organisms. There is a fairly rigid cell wall made up of unusual chemical components. The nucleus consists of a single two-stranded molecule of deoxyribonucleic acid—DNA.

In 1884, Hans Christian Gram developed a differential staining method for bacteria that is still used today to classify these microorganisms. After initial staining with the dye methyl violet, Gram treated bacteria with iodine solution and then with ethyl alcohol. Some bacteria retained the initial color when treated with alcohol; others did not. Those that retain the color are "gram-positive"; those that do not are "gram-negative."

Cells of yeasts and fungi are remarkably similar to the cells of animals and higher plants. There are mitochondria for energy metabolism during aerobic processes, ribosomes for protein synthesis, and chromosomes for transmission of genetic information.

Fungi are a diverse group. They include mushrooms and toadstools, as well as microscopic organisms. Only the latter group is of interest in fermentation science. They may be divided into multicellular forms, such as molds, and unicellular forms, such as yeasts.

Actinomycetes are usually considered somewhat intermediate between bacteria and fungi. Although some of these organisms are pathogenic, many are beneficial. In soil, they are responsible for much of the conversion of organic matter. Streptomycetes are the source of a number of important antibiotics.

NUTRITIONAL REQUIREMENTS OF MICROORGANISMS

Any living cell, whether it exists in a microorganism, an animal, or a man, has nutritional needs. The similarity in essential dietary requirements of such diverse groups is striking. All cells need a source of energy and materials from which various cell components can be synthesized.

The source of energy for most fermentative microorganisms is a simple sugar, such as glucose. Other requirements are mineral salts, vitamins, and a source of nitrogen for synthesis of essential molecules, such as proteins and nucleic acid. Some microorganisms can use ammonium ion (NH_4^+) as a source of nitrogen; more commonly, amino acids are required for microorganisms that grow during fermentation. Some microorganisms need as many as 12 different amino acids for growth.

Inorganic requirements include such ions as potassium, magnesium, manganese, iron, phosphate, and sulfate. Phosphate is probably the most important because of its role in energy metabolism.

Many microorganisms require vitamins, especially those of the "B" group. Several of these vitamins are known to be a part of enzyme cofactors, but the function of others is still unknown. It is known only that they are needed for growth. Many yeasts require vitamin B1 (thiamine) when fermenting, but not when growing aerobically.

The raw materials used in making beer and wine almost always have enough of the nutrients required for completion of fermentation. A fermentation may stop because of a depleted energy source. If this happens, addition of more sugar usually revives the fermentation. With an adequate source of energy and sufficient nutrients, an alcoholic fermentation will continue until the alcohol content reaches a level at which it is toxic to the yeast, causing its death. About 12% to 15%, by volume, of alcohol is usually lethal to the yeast.

Aerobic and Anaerobic Microorganisms

As we saw in Chapter 2, the products of respiration in man, as well as in lower animals and plants, are carbon dioxide, water, and energy. These same products are produced by respiration in most microorganisms. Just as man uses oxygen in respiration, so do these microorganisms. The organisms which require oxygen, or air, are called aerobic organisms, or aerobes. Some microorganisms do not require oxygen; these are anaerobic organisms, or anaerobes. Indeed, some anaerobic organisms are killed by oxygen. These are "obligate anaerobes." Still another class exists, known as "facultative anaerobes." These are organisms which can live in the absence of oxygen, although they do not ordinarily do so.

Respiration is a more efficient path for generating energy than is fermentation. In respiration, carbohydrate is oxidized completely to carbon dioxide and water; in fermentation, it is only partially oxidized to carbon dioxide and water. The part of the carbohydrate that is converted to ethyl alcohol does not produce energy.

Pasteur defined fermentation as "life without air"; the classical biologist used the word to mean the enzymatic anaerobic breakdown of carbohydrates; the modern definition of fermentation must be much broader, since industrial processes now use both aerobic and anaerobic operations which are called fermentations. One worker has suggested that the definition should be "microbial action, which is controlled by man to make products useful to man."

ALCOHOLIC FERMENTATION

The most generally familiar fermentation, and the one we shall consider first, is alcohol fermentation—an anaerobic fermentation mediated by certain strains of yeast. Almost any carbohydrate-rich material can serve as the substrate in the process. In the beverage industries, barley, malt, corn, and rice are often used. Blackstrap molasses is frequently used for making industrial alcohol because it is inexpensive, easy to handle, and contains 50% fermentable carbohydrate. The substrate is usually diluted to a level of about 10% to 15% sugar, ammonium salts are added as a source of nitrogen, and the pH is adjusted to between 5 and 6. As in any fermentation, the medium must be sterile before the desired microorganism is added. Thus, the medium is pasteurized, cooled, and inoculated with the yeast Saccharomyces cerevisiae.

Alcoholic fermentations are usually carried out at rather low temperatures, between 7° and 15°C, depending on the mode of fermentation and the desired final product. Since the fermentation generates considerable heat, it is necessary to use cooling coils to maintain a constant low temperature. The process is usually carried out for 1 to 2 weeks when alcoholic beverages are made.

When grain is used as a substrate for fermentation, it is necessary to convert the starch to sugar, since yeasts cannot ferment starch. The grain is ground and heated in an aqueous gravy to dissolve the starch. It is cooled to 145° F and treated with malt (partially germinated barley), which is a good source of starch-degrading enzymes, especially beta amylase. The starch is converted to maltose—a sugar fermentable by yeast. The "beer" produced by grain fermentations usually contains about 5% to 8% alcohol by volume. The alcohol is isolated by distillation.

The alcohol fermentation is used in the production of alcoholic beverages and in the baking of bread. In the former case, the production of alcohol is important; in the latter, the carbon dioxide is the useful product. The gas produced by the fermentation riddles the dough with small holes and causes it to rise. In baking, the gas expands and boils off, leaving the bread light and spongy. Uses of enzymes in the baking and beverage industries are considered further in Chapter 6.

The overall reaction occurring in an alcohol fermentation is summarized by the equation:

$$C_6H_{12}O_6 \xrightarrow{\text{Yeast}} 2\ C_2H_5OH \text{ and } 2\ CO_2 \text{ and } 31{,}200 \text{ Calories}$$

$$\text{Glucose} \longrightarrow \text{Ethyl alcohol and Carbon dioxide and Energy}$$

The enzymic mechanism, by which these products are formed, is called the Embden-Meyerhof-Parnas glycolytic pathway (see Figure 2–2). It is a 12-step sequence with each step catalyzed by a different enzyme.

Lactic Acid Fermentations

The undesirable microorganisms, indicted by Pasteur as causes of the souring of wine, were lactic acid bacteria. These microorganisms are a frequent contaminant in yeast fermentations. They occur in two varieties: homofermentative lactic acid bacteria convert glucose to lactic acid only; the heterofermentative variety uses glucose to form lactic acid, ethanol, and carbon dioxide.

The homofermentative bacteria use the first 10 steps of the Embden-Meyerhof-Parnas glycolytic pathway, but they do not have the enzymes needed for the last two steps. One of the missing enzymes is responsible for liberating carbon dioxide. This explains why there is no carbon dioxide released by homofermentative bacteria.

This glycolytic pathway used by homofermentative bacteria is the same path used by a human muscle when it is overworked. The accumulation of lactic acid in the muscle is a major source of fatigue.

Sauerkraut and dill pickles are made by processes which use heterofermentative lactic acid organisms. Cabbage and cucumbers are hosts to these organisms, as well as to less desirable bacteria. The growth of the lactic acid producers is promoted by steeping the vegetables in a salt solution.

Solvent-Producing Fermentations

Acetone was needed for the production of explosives in World War I, and a fermentation was found capable of yielding this product. The gram-positive bacteria Clostridium acetobutylicum fermented starch giving acetone, butyl alcohol, and ethyl alcohol in a 3:6:1 ratio. Corn mashes of 10%, which contained about 70% starch, were used as substrate and each 100 grams of starch gave about 30 grams of mixed solvents.

With the end of the war, the need for acetone declined, but butyl alcohol became very important as a component of the solvents used for automobile lacquers. Nonbiological synthetic processes became economically competitive with solvent-producing fermentation methods during the early years of the depression and ways were sought to make the fermentation process cheaper.

Strains of Clostridium were found that could ferment molasses—a cheaper substrate. The fermentation was more rapid, and the product contained approximately 70% butyl alcohol.

Chemical synthesis has continued to compete with fermentation as a route to simple molecules and the vast petroleum industry has supplied most of the solvents used in recent years. With the onset of the alleged oil shortage, prices for solvents produced by the petroleum industry have increased as dramatically as oil company profits. A revival of the fermentative production of solvents may be expected to make prices more reasonable.

AEROBIC FERMENTATIONS

The economic competition from the chemical and oil industries caused fermentation scientists to look for new worlds to conquer. The discovery of penicillin and the consequent search for other antibiotics was a boon to the fermentation industry. These complex materials are made easily by fermentation, but chemical synthesis would be prohibitively difficult. Thus, the fermentation industry is now producing products which are not likely to be produced by competing industries. In addition to antibiotics, steroids, vitamins, and enzymes are produced by microorganisms. One of the developments which made it possible to prepare these complex molecules was the discovery of aerobic fermentations.

ANTIBIOTICS

After a week away from the laboratory, the Scottish bacteriologist, Alexander Fleming, discovered that a patch of green mold had contaminated one of his cultures. The mold was interesting because it had antibacterial action. The green mold was Penicillium notatum, and Fleming called the antibacterial product, penicillin. This first antibiotic compound acts on gram-negative bacilli (bacteria) by interfering with the activity of particular enzyme systems involved in bacterial growth. It is effective against staphylococci, streptococci, meningococci, and other pathogenic organisms. Dr. Fleming's life was prolonged when penicillin saved him from a serious attack of pneumonia. Numerous other lives have been saved by this antibiotic, including many wounded soldiers in World War II.

There are six closely related natural forms of penicillin and chemists have used these natural forms to prepare hundreds of semisynthetic forms. These are discussed in Chapter 14.

Penicillin is produced industrially by a technique known as "submerged aerobic fermentation." In this method, large volumes of sterile air (produced by passing compressed air through sterile filters packed with glass wool or carbon) are continuously passed into the medium with mechanical agitation. The method produces more rapid fermentations than alternative procedures. It is conveniently done on a large scale, and it requires little manpower.

Penicillium chrysogenum is the usual mold used. The sterilized medium contains corn steep liquor, glucose, lactose, sodium nitrate, magnesium and zinc sulfates, phenethylamine, and calcium carbonate. Fermentation times vary from 60 to 200 hours.

Penicillin therapy sometimes produces severe toxic reactions. Moreover, microorganisms develop a penicillin resistance. These facts provided some of the motivation to find more effective antibiotics.

Streptomycin was the second antibiotic used therapeutically. Discovered by Waksman at Rutgers in 1944, this compound was active against gram-negative bacteria and against Mycobacterium tuberculosis —the bacterium that causes tuberculosis. It is also used in the treatment of plant diseases.

Streptomycin was the first useful antibiotic obtained from Actinomycetes—a group of organisms which have subsequently produced many more antibiotics. Streptomyces griseus are cultivated by submerged aerobic fermentation at 27° C on a medium containing carbohydrate, soybean meal, and inorganic salts.

Chloramphenicol was the first "broad-spectrum" antibiotic—so called because it is effective against a broad spectrum of diseases caused by gram-positive and gram-negative bacteria as well as other pathogens. Since it generally is prepared by chemical synthesis, we shall go on to a group of broad-spectrum antibiotics produced by fermentation—the tetracyclines.

These include aureomycin, teramycin, and many related broad-spectrum antibiotics. They are useful in treating diseases in man, plants, and animals. In addition, they have been found to increase food utilization and improve marketable weight in livestock.

Many varieties of the streptomyces produce tetracyclines. The media used vary considerably, but a typical medium may contain corn steep liquor, sucrose, an ammonium salt, and various combinations of metal ions, usually including sulfates and phosphates. Submerged aerobic fermentation is used.

Steroid Conversions

Steroids are complex organic molecules that have a number of important functions in the chemistry of life. Cholesterol—the most frequently cussed and discussed steroid—is the most predominant steroid in man, and it is a key intermediate in the biosynthesis of other steroids. A number of hormones produced by the adrenal glands and the sex organs are steroids. The bile salts, which are important in digestion of fats, contain steroids.

In medicine, steroids such as cortisone are used to treat rheumatoid arthritis, allergies, skin and eye diseases. They have been used as antifertility drugs and have found some success in the treatment of cancer.

Steroids are generally obtained from plant sources or synthesized by laborious chemical methods. Although microorganisms are not used for the preparation of these molecules, they are useful in modifying them. Quite frequently, the plant or synthetic steroid needs an alcohol (OH) group or a double bond at a specific position in the molecule. Such seemingly minor changes can be very difficult to bring about by ordinary chemical means. They can effect quite drastic changes in the biological activity of the molecule. It is often possible to find a microorganism to make the desired conversion. For instance, over 500 different microorganisms are able to convert progesterone to 11–alpha–hydroxyprogesterone by adding an OH group at a specific position. This conversion gives a compound related to cortisone.

To determine if an organism will cause a desired transformation, the organism is grown in a suitable medium for a few days; the steroid, dissolved in a suitable, water-miscible solvent, is added and the mixture is incubated for 1 or 2 days. The product is isolated and purified and its structure is determined. If the transformation was effected, the process can be developed.

Steroid transformations are carried out in submerged culture tanks. The steroids are usually added to the growing microorganisms near the end of the organism's growth phase. Transformations are structurally minor, but quite important in effect. That transformations occur at all is somewhat surprising in view of the fact that steroids are essentially insoluble in water.

AMINO ACIDS

Amino acids, as we have seen, are the building blocks of proteins. They are used as supplements in human foods and animal feeds. They also have been used as medicines and cosmetics. Japanese scientists have used them as chemical intermediates and agricultural chemicals.

Japan experienced a food crisis during World War II and the early postwar years, and this crisis led to the preeminence of the Japanese in the areas of synthetic protein and amino acid production. Not blessed with abundant, fertile farm land, they have exploited other means for the production of proteins and amino acids. One of these means is fermentation.

Probably the most widely used amino acid is glutamic acid. It is most frequently used as the salt, monosodium glutamate, which is sold in the United States under the trade name "Accent." Kinoshita has described a glutamic acid fermentation using the bacterium Micrococcus glutamicus. The organism is grown at 28° C for 24 hours in flasks on rotary shakers moving at 220 rpm. The broth contains glucose, peptone, meat extract, and sodium chloride. The pH is maintained at 7.0 to 7.2. A 10% inoculum of this growth is added to a sterile fermentation medium containing ammonium, iron, magnesium, manganese, sulfate, and phosphate ions and glucose. Biotin (a B vitamin) is also used. The fermentation is carried out at 25° C on a rotary shaker with periodic addition of urea to maintain pH.

The fermentation is described in a patent and the process, as described, is a laboratory-scale preparation, which gives about 50 grams of product per liter of broth. Since this amino acid is produced in quantities of hundreds of thousands of tons per year, it is obvious that large-scale fermentations are available. It is a frequent practice in the fermentation industry to describe the laboratory preparation to gain patent protection, but to keep the industrial process secret.

Cereal grains are frequently deficient in the essential amino acids: lysine, tryptophan, threonine, and methionine. Plant scientists are busily trying to develop cereals richer in these nutrients. Meanwhile, the amino acids can be prepared by chemical synthesis, or by fermentation, for use as a nutritional supplement. Fermentation has the advantage of producing only the useful L–form, while chemical synthesis produces equal quantities of D– and L–forms. Methods for converting the D–form to the L–form by use of enzymes are described in Chapter 6.

Fermentation methods are available for all of the important

amino acids. Only glutamic acid and lysine are produced on a large scale by fermentation in the United States.

PRODUCTION OF ENZYMES

Enzymes may be obtained from animals, plants, or microorganisms. The first two sources are severely limited; the last is virtually unlimited. Microorganisms can be used to meet all possible commercial needs for enzymes. They also make available a number of enzymes which have not been obtained from plants or animals. These include steroid-converting enzymes, heat-stable amylases, therapeutic enzymes, such as streptokinase-streptodornase, and others.

All currently used enzyme-producing fermentations are aerobic processes. Two basic methods are used: the semisolid culture method, developed by Takamine in the early years of the twentieth century, and the more recently developed submerged culture method. In the semisolid culture method, the microorganisms grow on a moist, solid substrate, such as wheat bran. The organism is grown in liquid cultures in large tanks, when the submerged culture method is used. The latter method is preferred because of its flexibility, but Takamine's method is still in use.

When the semisolid culture method is used, wheat bran is mixed with other nutrients, which may include carbohydrate, ammonium salts, growth stimulants, and minerals. Buffer is added to this mash and it is sterilized. After cooling, the organism is added and the inoculated bran is spread in thin layers and incubated in chambers or in slowly rotating drums. Temperature, humidity, and aeration are controlled by circulating cool, moist, conditioned air over the mixture. Air is able to penetrate the porous bran and the organism grows throughout.

The submerged culture method employs closed fermentors, which vary in capacity from 1,000 gallons to tens of thousands of gallons. Liquid medium is usually sterilized on a continuous basis and piped into the sterile fermentor. The medium is cooled and the organism added. The system is continuously and mechanically agitated and sterile air is supplied throughout the fermentation.

Whichever procedure is used, the enzyme activity is determined frequently enough to permit harvesting the product when the desired activity is maximal. This is usually within 1 to 7 days.

Some enzymes are secreted by the microorganisms into the surrounding fluid. Most microbial enzymes of industrial importance are of this type. The first step in recovering such enzymes from the reaction mixture is to filter away all solid material including cell debris. The solution is then chilled and the enzymes are precipitated by addition of a water-soluble organic solvent, such as acetone or an alcohol, or by a salt, such as ammonium sulfate. The precipitated mass can then be purified to the desired extent by standard physical and chemical techniques.

If the enzyme is an intracellular one (not excreted into the medium), it is necessary to break the cells, usually by mechanical disruption (as the Buchners did with yeast cells). The enzyme then goes into solution, the cell debris is filtered off, and the enzyme is precipitated.

PRODUCTION OF OTHER PROTEIN BY FERMENTATION

Under anaerobic conditions, yeasts produce ethyl alcohol, but, in the presence of excess oxygen, the alcohol is further decomposed to carbon dioxide and water. Under these aerobic conditions, there is an increase in the yeast crop. This yeast-producing fermentation is a useful means for obtaining protein and B vitamins to be used in enriching foods and animal feeds. The yeast cells are isolated, heat-killed, and dried. In the dry form, they provide a high quality dietary supplement.

But these microorganisms use carbohydrate as nourishment. In recent years, it has been found possible to grow microorganisms on materials which are not useful in human nutrition, most notably, hydrocarbons. The yeasts, which propagate on this substrate, have a cell wall which resists enzymic digestion. This has been one of the limitations of this source of "single-cell protein." Harada and co-workers in Japan have described a chemical method for causing a mutation in a strain of yeasts—a mutation which makes the yeasts readily susceptible to digestion. They believe that the mutation caused a defect in the cell wall, which permits digestive enzymes to get into the cell. If the mutation is too severe, the organisms will not proliferate; by carefully limiting the extent of the mutation, a yeast is obtained that reproduces on hydrocarbon stock and that can be digested by pepsin in hydrochloric acid. The amino acid content of the normal parent cell and the mutant daughter are essentially the same. Protein is easily isolated from the mutant strain by use of a mechanical homogenizer; the parent cell resists this treatment. Both the parent and the mutant are capable of propagating in a hydrocarbon medium. The quantity of cell mass formed is equivalent to the mass of hydrocarbon consumed.

There are still some technical problems in the development of single-cell protein; but, for a world threatened with the likelihood of starvation, the discovery of microorganisms which can produce protein from hydrocarbons is cause for hope.

Chapter Five

Industrial Uses of Enzymes

PROBABLY THE BEST-KNOWN and most widely used industrial application of enzymes is in detergents and presoaks. Although literature on enzyme detergents appeared as early as 1915, it was not until 1963 that the first successful enzyme detergent presoak was sold, in Holland. By 1967, European detergent makers introduced sodium perborate detergents containing enzymes. This was no longer just a presoak, but a complete wash cycle detergent. Tide XK, the first enzyme detergent sold in the United States, was available in test market areas for the first time in 1966.

Enzyme-containing laundry products before 1960 were based on animal proteases, which were not suitable for use under laundry conditions. The availability of commercial quantities of stable bacterial proteases was a breakthrough which led to commercial enzyme detergents. Microorganisms known as Bacillus subtilis are used to produce laundry enzymes by fermentation procedures.

Proteases and alpha amylase are usually blended directly with a dry detergent base. More stable enzymes can be dissolved or suspended in water and sprayed onto a base material, which is then mixed with the detergent base. Enzymes are usually added to the dry detergent in the last step before packaging.

Presoaks are used to loosen or remove heavy stains or hard-to-remove stains before laundering with a regular detergent. Presoaking times are varied from a few minutes to several hours, based on the difficulty of removal of the stain.

Enzyme detergents are, indeed, more effective than pure detergents in removing "difficult" stains, such as blood, grass, and the like. For complete removal of such stains, long-term contact with the enzyme, as in a presoak, is often needed.

As in any other chemical reaction, temperature is an important consideration. The typical laundry proteases are most useful at wash temperatures around 120° F. Cold-water washing reduces the enzyme activity and temperatures above 120° F, in the presence of detergent, denatures the proteases.

The optimum pH range for the laundry detergent enzymes is 8.5 to 10.5. Most American detergents have pH values in this range.

Safety

Early tests indicated that the use of enzymes in laundry products was a completely safe, as well as highly effective, operation. In 1969, reports appeared which contradicted these earlier data. Adverse effects were seen on the skin as a result of handling the enzyme detergents and in the lungs due to inhalation of the enzyme-containing dust. As these findings became widespread, the popularity of enzymes in laundry products declined and many products were taken off the market or reformulated without enzymes.

Enzyme detergents should not be placed in the category of "a good idea that just didn't work." Research in this area is very abundant and many new patents are being granted for improved enzyme detergent formulations.

Dry Cleaning

Enzymes have been used for at least 30 years in the dry cleaning industry for spot removal. The usual dry cleaning solvents cannot remove proteinaceous stains, such as blood, grass, or eggs from clothing. Proteolytic enzymes are able to digest these stains without harming the fabric. Lipases and amylases have been similarly used.

ENZYMES IN DRAIN CLEANING,
SEWAGE TREATMENT, AND POLLUTION CONTROL

Waste materials in drains and sewage systems are composed primarily of organic materials, such as those found in foods—carbohydrates, fats, proteins, and fiber. Natural disposal of these organic wastes depends on bacteria. As we have seen, bacterial action is due to enzymic activity.

Miles Laboratories produces a drain cleaning and sewage treatment mixture called Sanizyme, which consists of aerobic (oxygen-using) and anaerobic bacteria, along with extra enzymes including amylase, protease, lipase, and cellulase. The amylase acts on carbohydrates; protease acts on proteins; lipase acts on fatty materials; and cellulase acts on cellulose-containing products, such as paper and vegetable materials. These enzymes accelerate the growth and multiplication of the bacteria and increase their effectiveness.

"Activated sludge systems," composed of various microorganisms, are useful in treating industrial waste waters. It is often necessary to use pretreatments and "polishing" treatments in conjunction with the biological system.

Microorganisms and enzymes have been useful tools in the fight against water pollution. They promise to be much more useful in the future. With our alleged oil shortage, and our very real food shortage, scientists are seeking ways to convert waste materials to energy-producing products and foodstuffs. Quite naturally, they have turned to microorganisms and enzymes for these conversions.

It has been estimated that a billion tons of solid organic waste are generated in the United States in a year. Of this, nearly 200 million tons are animal manure. The rest includes agricultural and food wastes, urban refuse, industrial wastes, and municipal sewage. These materials can be converted to food or energy by use of microorganisms.

Produce Food from Waste Materials

Swedish scientists have shown that it is possible to combine waste disposal with a profitable process which provides high-grade protein. They have grown yeasts on industrial wastes. The yeast obtained is of high quality, as judged by digestibility and by vitamin and amino acid content. The amino acid content is similar to that of soybeans or fish. It is high in lysine—the essential amino acid frequently low in cereals. The yeast is a useful supplement for human foods and animal feeds.

At the same time, the elimination of waste materials in the process is comparable with that provided by the usual activated sludge and biological filtration processes. The yeast process is also cheaper and more efficient in removing nitrogen and phosphorus-containing pollutants.

Produce Energy from Waste Materials

A great many organic compounds can be converted to methane by microorganisms with a theoretical fuel energy recovery of more than 90%. The methane is contaminated only with carbon dioxide and trace amounts of other easily removable molecules. Methane can be used directly in equipment designed for natural gas. It is a very clean and convenient fuel. The equipment needed is modest—a simple fermentation vessel, a gas-collecting device, and gas-storage vessels. The process can be carried out on any convenient scale, depending on the amount of organic material available. This is nearly limitless.

ENZYMES IN THE LEATHER INDUSTRY

Animal hides are converted to leather by a process called tanning, which involves soaking the hides in a chemical bath. There are a number of steps between the removal of the hide and its conversion to leather.

The hide is cured by use of dry salt or brine. This reduces the moisture content of the hides from 70% to about 40%. About 15% salt is added. This preserves the hide until it is tanned.

Pretanning is carried out in a building called a "beamhouse," so the process is called the "beamhouse operation." The first part of this involves soaking the hides. This removes some of the salt so it will not interfere with subsequent steps.

This is the first step of the process in which enzymes have been used. Proteolytic enzymes decompose hide proteins other than collagen, the most abundant and most important protein of leather. The removal of these proteins makes the product soft and pliable. The soaking of skins having a high level of fatty materials can be facilitated by including a lipase in the soaking solution, along with the protease.

In the liming process which follows, the hide is treated with a mixture of hydrated lime and a sulfide for several days. This facilitates the process of removing the hair from the skin. The effluent from the liming process presents a serious pollution problem. Alternative methods of removing the hair were sought to reduce this problem. The answer was provided by an ancient process known as sweating. This process consists of letting the soaked hides hang for several days under controlled conditions of temperature and humidity until the epidermis (skin) and the hair follicles are attacked by bacteria and the hair is loosened. It was reasoned that, since these bacteria exert their effect by enzyme action, the direct application of enzymes might be useful in removing the hair.

Proteolytic enzymes, derived from plant, animal, or microbial sources, have been found quite efficient in unhairing hides. The finished leathers compare favorably with leathers produced by the liming process. The enzyme process is simpler and it reduces the pollution problem. Both the pelt and the hair or wool can be removed in good condition.

The final steps before tanning are bating and degreasing. Bating is essential in the manufacture of soft leather having characteristic grain and superior quality. The main object of bating is the removal of some of the nonleather-forming proteins of the hide and the opening of the fibers to make it easy for the tanning agents to penetrate. The materials classically used by tanners for bating were dog and pigeon manures. It was discovered that proteolytic enzymes were the active components of the manure, and this discovery has led to a more aesthetically pleasing process—the use of enzymes instead of manure for bating.

If the hide is to be stored for an extended period before tanning, it is usually pickled in an acid salt solution. If it is not to be stored, it is usually degreased by treating it with a fungal or bacterial lipase. If not removed, natural greases can cause a number of defects in the finished leather.

After these stages of pretreatment, the hide is treated with a solution of the tanning agent and is thereby converted to the product we know as leather.

COSMETIC USES OF ENZYMES

The patent literature provides abundant examples of cosmetic applications of enzymes. This literature tends to feature exorbitant claims designed to impress the patent examiner, so it must be considered with the proverbial grain of salt. With this caveat, we shall mention a few examples of cosmetic applications.

A German patent described an enzyme-containing preparation which was said to be suitable for local treatment of cellulitis, rashes, sunburn, and blisters. It was further said to be effective as an antirheumatic, to have anti-inflammatory properties, to promote tissue regeneration, and to be useful in treating varicose veins. The enzyme was extracted from the skin covering or navel cord tissues of cattle or sheep fetuses. Preparations intended for use as anti-inflammatory agents may also contain additional enzymes such as hyaluronidase, lipase, and thiomucase.

Another German patent describes a stable cream for defatting human tissues. This preparation contains the enzyme, lipase, and stabilizers and emulsifiers in an ointment base. Ox bile is an optional activator. Hyaluronidase and thiomucase are sometimes added to promote diffusion of the lipase. The stabilized creams were fully active after 24 months storage, whereas those containing no stabilizer had no lipase activity after 2 or 3 weeks of storage.

Just as enzymes are useful in unhairing hides for leather manufacture, it seems that they may also be useful for removing hair from milady's legs. Depilatories containing the enzyme, keratinase, as the active ingredient, have been described in United States patents. The widely used malodorous chemical preparations usually contain a thioglycolate or similar sulfur-containing compound. These materials are used at a very high pH (about 8.5 to 9.5), which is not very gentle on the skin. The enzyme depilatory is active at pH 7.5, close to the physiological pH. The enzyme product is not as rapid as the chemical one, but it can be left on the skin overnight without harm. It can be perfumed very easily, but there is no objectionable odor to mask.

Enzymes in the Textile Industry

In order to bind threads together and prevent their breaking in the loom, the cloth manufacturer coats his threads with an adhesive material called "size" or "sizing." By protecting the threads from the effects of mechanical stresses and friction, the sizing makes it possible to weave the cloth more efficiently.

But, when the finisher receives the woven cloth, he must remove the sizing completely from the cloth to effect uniform bleaching, dyeing, printing, or finishing. This removal of sizing is the principal use of enzymes in the textile industry.

The most frequently used sizing material is starch. Lubricants, softeners, and binders are frequently included in the sizing preparation. Proteins, gums, and waxy materials are sometimes used in sizing.

Since starch is the primary sizing agent, a starch-degrading enzyme, an alpha amylase, is the primary desizing agent. This enzyme rapidly attacks starch molecules at random points, breaking the molecule into relatively small fragments called dextrins.

Starch used in sizing contains approximately 80% amylopectin and 20% amylose. These structures are illustrated in Figure 5–1.

Figure 5–1   Structures of starch components.

Amylose fraction of starch containing only 1,4-bonds and therefore a linear chain:

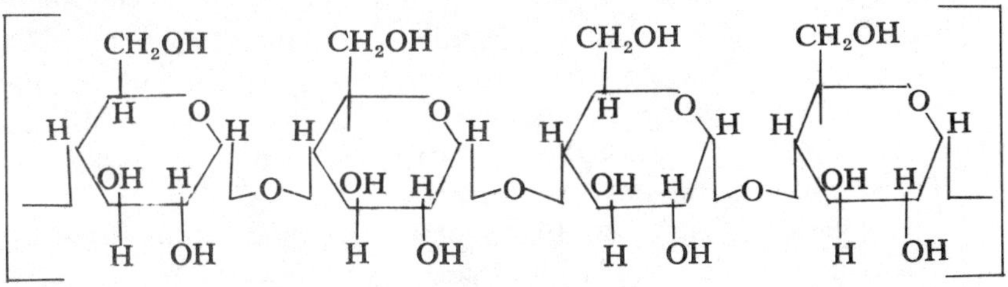

Amylopectin fraction of starch containing 1,6-bonds, representing branch points:

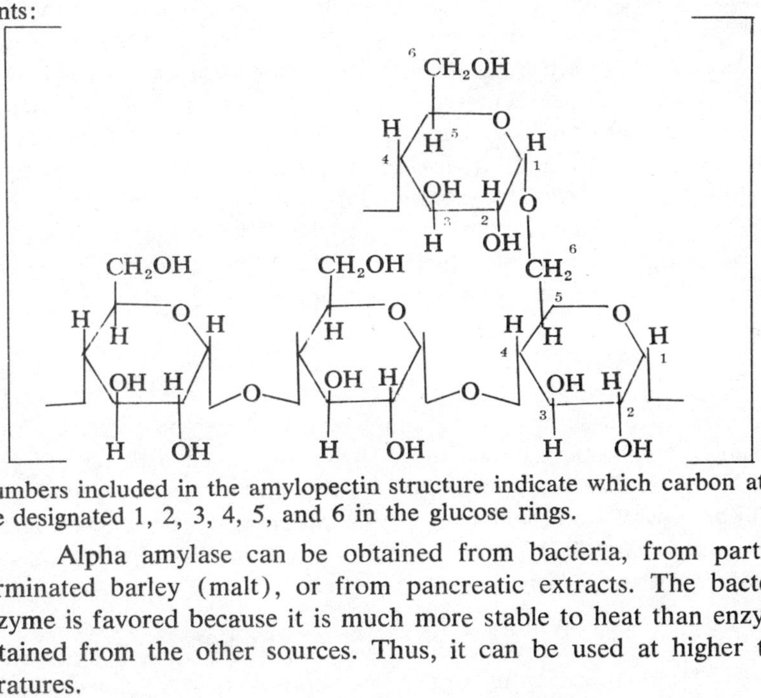

Numbers included in the amylopectin structure indicate which carbon atoms are designated 1, 2, 3, 4, 5, and 6 in the glucose rings.

Alpha amylase can be obtained from bacteria, from partially germinated barley (malt), or from pancreatic extracts. The bacterial enzyme is favored because it is much more stable to heat than enzymes obtained from the other sources. Thus, it can be used at higher temperatures.

The use of elevated temperatures is desirable for a number of reasons. It permits rapid, continuous desizing and it predisposes the starch for enzymic decomposition by gelatinizing it. Nongelatinized starch resists enzymic attack. Higher temperatures also help in liquefying waxy or fatty lubricants, which are used in sizing mixtures. This facilitates the penetration of the enzyme into the fibers, so that sizing is removed from the inside of the fibers, not merely from the surface.

The enzyme action decomposes starch into smaller, more soluble units, but this is not enough. Those smaller units must be thoroughly removed from the cloth. If enzyme action is continued long enough, the decomposition products are removable by cold-water washing. If cold-water washing is inadequate, washing at 175° F is usually sufficient. More resistant sizing is washed out by treatment with 1% caustic solution at 180° F, after the enzymic decomposition.

Proteins and gums, used in association with starch in sizing, can form films around the starch, preventing alpha amylase attack. Industrial bacterial amylase preparations are contaminated with proteases and hemicellulases which can decompose these materials. These enzymes are not as heat resistant as the amylase, so the treatment is started at 45° F when this enzymic activity is desired. After the protease and hemicellulase have decomposed their substrates, the temperature is raised to accelerate alpha amylase activity.

Sometimes the size is entirely protein and gum. This is particularly true of acetate and continuous filament viscose rayon cloths. Although gelatin, the most frequently used protein size, is soluble, its complete removal is facilitated by proteases. It is important to completely remove protein, whether it is the principal size or an adjunct, since protein will attract dirt and bind it to the fabric.

Enzymes in Photography

As the cost of silver continues to rise, the recovery of this precious metal from scrap photographic film becomes progressively more desirable. A commercial enzyme preparation called Takamine Gelatinase No. 3 (Miles Laboratories) at a concentration of 1½ pounds in 6,700 pounds of water will strip the gelatin emulsion from 1 ton of film in an hour at 60° C. With the gelatin dissolved away, the silver salts are readily recovered.

Immobilized Enzymes in Industry

Many industrial processes, particularly in the pharmaceutical industry, are closely guarded secrets. Many people suspect that the pharmaceutical industry makes use of immobilized enzymes in producing some of their complex products, but no publications have as yet confirmed this belief.

It seems likely that, by the time this book appears, commercial processes for making sugar from starch, by use of immobilized gluco-amylase, will be operational. At the time of this writing (early 1976), there does not seem to be any industrial process using immobilized enzymes in the United States. As we shall see in Chapter 6, the Japanese have taken the lead in the commercial use of these catalysts.

Chapter Six

Enzymes in the Food Industry

IN CONSIDERING FOOD-RELATED enzymes, it is appropriate to begin with enzymes related to the production of amino acids—the fundamental building blocks of proteins. The greatest need of the world's starving millions is for increased amino acids in the diet. These nutrients are normally obtained from ingested protein foods such as meat, eggs, fish, dairy products, cereals, and beans, but modern techniques make it possible to enrich any food with amino acids.

Chemical techniques for synthesizing amino acids have one serious flaw. All such techniques result in racemic mixtures—equal quantities of the D– and L–forms. Since only the L–form can be metabolized and used for protein synthesis in the body, the mixture must be separated. Physical and chemical methods for separating such mixtures are both difficult and inefficient.

Since 1954, workers at the Tanabe Seiyaku Co., Ltd., of Osaka, Japan, have used an enzyme—aminoacylase, derived from a mold—for separating such mixtures. The materials synthesized chemically are acyl amino acids—derivatives of the desired products. The enzyme releases the desired L–form, leaving the D–form unchanged. Using acyl–D, L–phenylalanine as a typical substrate, the enzyme reaction is:

Equation 6-1

$$\text{D,L-}\;\underset{\text{(phenyl)}}{\overset{\overset{\displaystyle NHCOR}{|}}{CH_2CH\text{–}COOH}}\;\text{ and } H_2O \xrightarrow[\text{-----}]{\text{Aminoacylase}} \text{L-}\;\underset{\text{(phenyl)}}{\overset{\overset{\displaystyle NH_2}{|}}{CH_2CH\text{–}COOH}}\;\text{ and } RCOOH$$

$$\text{D-}\;\underset{\text{(phenyl)}}{\overset{\overset{\displaystyle NHCOR}{|}}{CH_2CH\text{–}COOH}}$$

N–acyl–D,L–phenylalanine and Water

$$\xrightarrow[\text{-----}]{\text{Aminoacylase}}$$

L–phenylalanine and Carboxylic acid
N–acyl–D–phenylalanine

The two compounds are easily separated after such treatment because of differences in their solubilities.

Although it is difficult to separate D– and L–forms from racemic mixtures (except by enzyme action), the conversion of one form to a racemic mixture is easy. The acyl–D–amino acid which is separated from the nutritional L–amino acid can be converted to a racemic mixture and put through the enzyme reactor to obtain additional L–amino acid. In this manner, all of the amino acid is eventually converted.

Aminoacylase has absolute specificity for the L–amino acid derivative. It effects complete conversion of the acyl–L–amino acid, leaving the D–isomer intact.

While the aminoacylase process was an important means for isolating essential nutrients, it had a number of limitations. The soluble enzyme was necessarily used in a "batch" process instead of a "continuous" one. The stability of the enzyme was low and it could not be recovered and reused. Control of reaction conditions and automation was difficult and the process gave low yields and impure products.

Interest in immobilized enzymes increased dramatically in the 1960s, and the aminoacylase system was studied in depth by the Japanese scientists. They found that adsorption of the enzyme onto a cellulose-like material having ionic groups grafted onto the molecule was a useful method for immobilizing aminoacylase. The support was diethyl-aminoethyl-Sephadex (DEAES).

A crude enzyme preparation was used and the aminoacylase was completely adsorbed. Some enzyme purification apparently was achieved in the adsorption process for the enzymic activity of the DEAES-aminoacylase adduct was 134% of that of the soluble enzyme.

Low levels of cobalt ion are used to activate the adsorbed enzyme or the soluble one. The adsorbed enzyme shows maximum activity at 62° C. Optimum pH is different for the various amino acids, and the pH in a given packed bed reactor is the pH found optimum for the conversion of the amino acid treated in that reactor.

The adsorbed enzyme retained the absolute specificity of the native enzyme, completely converting the L–isomer while leaving the D–form intact. Aminoacylase remained firmly attached to the support as long as dilute substrate solutions were used. It could be leached from the column by more concentrated substrate solutions (concentrations greater than 0.2 molar).

After 17 days of continuous operation, the column retained 60% of its initial activity. Deteriorated enzyme columns can be regenerated by simply adding fresh aminoacylase to restore the original level of activity.

The process using immobilized aminoacylase is easily controlled and has been automated. It produces 100% yields of pure products. Overall production costs of the amino acids produced by the immobilized enzyme process are about 60% of the cost of the batch process using soluble enzyme. The saving is due to a remarkable reduction of amounts of substrate and enzyme needed and to reduced labor costs.

The DEAES–aminoacylase process was put into industrial production in Japan in 1969 and was the world's first industrial use of immobilized enzymes.

Another amino acid producing process was studied carefully by the Tanabe Seiyaku scientists—the production of aspartic acid from fumaric acid and ammonia catalyzed by the enzyme aspartase.

Aspartic acid, which is widely used in medicines and as a food additive, has been prepared by fermentation and by the action of free aspartase on fumaric acid and ammonia. This process had the same drawbacks as the soluble aminoacylase process. The enzyme was immobilized by entrapping it in a polyacrylamide gel. The activity of the entrapped enzyme and its stability under operating conditions were dis-

Equation 6–2

$$\text{HO–C(=O)–CH(H)=CH(H)–C(=O)OH} \quad \text{and} \quad NH_3 \xrightarrow{\text{Aspartase}} \text{HO–C(=O)–CH}_2\text{CH(NH}_2\text{)–C(=O)OH}$$

Fumaric acid and Ammonia $\xrightarrow{\text{Aspartase}}$ Aspartic acid

appointing. In addition, the enzyme had to be freed from inside the cells of its bacterial source, Escherichia coli—and added inconvenience and expense. Immobilized aspartase was not deemed industrially useful for aspartic acid production.

The Japanese scientists next considered the immobilization of the whole E. coli cell. Various immobilizing methods were tried and polyacrylamide entrapment gave the most active preparations. The entrapped cells were incubated in substrate solution for a day or two and an unexpected result was seen. The aspartase activity of the preparations increased almost tenfold. Apparently, the polyacrylamide treatment caused a splitting of the cells, making it easier for enzyme and substrate to come together. The large enzyme molecule cannot get out of the polyacrylamide trap, but the much smaller substrate molecules can diffuse in and out readily.

Entrapped cells had a pH optimum of 8.5 for aspartase activity, the same as the soluble enzyme, although the optimum for such activity in free whole cells was 10.5. Optimum temperature was 50° C. Manganese activated native and immobilized aspartase but not the intact or immobilized whole cells. Metal ions did have a protecting effect on the cells, however.

An automatically controlled reactor system, essentially the same as the immobilized enzyme reactor, was designed using polyacrylamide-entrapped E. coli cells as a source of aspartase activity. The column was operated at 37° C and had an operational half-life of 120 days. (After 120 days of operation, half of the original aspartase activity remained.) This was the first industrial application of immobilized whole cells in the world, and it reduced the overall cost of aspartic acid to about 60% of the cost of the batch process using nonimmobilized catalyst. The process has been operated industrially since 1973. Thus, a single Japanese company was the first to bring to fruition the industrial utilization of both immobilized enzymes and immobilized whole cells.

TENDERIZING AND FLAVORING MEATS

Natives of tropical countries rub a slice of the green fruit of the papaya tree over tough meat or dip the meat into papaya juice before cooking. Alternatively, they wrap the meat in papaya leaves overnight before cooking. Both of these processes make use of the proteolytic enzyme, papain, which is present in the juice of the fruit as well as in the latex of the tree and in the leaves. The American homemaker uses this same enzyme when she sprinkles meat tenderizer on her round steak.

The natural means of tenderizing meat is by aging. The meat is stored at 34° F for a time sufficient to permit natural enzymes and bacteria to break down the meat fibers. This process is limited not only by the risk of undesirable bacterial growth and excessive deterioration of the meat, but also by the requirement of maintaining large storage rooms at low temperatures for a long time.

Many advances in the use of proteolytic enzymes as meat tenderizers have been described in patents. Enzymes most frequently used are papain from the latex of the papaya tree, bromelin (bromelain) obtained from pineapple stems, and ficin from the latex of the fig tree. These plant proteases have considerable activity on connective tissue proteins, principally collagen and elastin, and have less activity on muscle fibers. Microbial proteases from the fungus Aspergillus oryzae, and the bacterium Bacillus subtilis are also used in meat tenderizing. They complement the plant proteases in that their primary action is on muscle tissues. They have only a slight effect on collagen and none on elastin. Combinations of plant and microbial proteases are sometimes used.

The meat industry has studied the use of pre- and postmortem injections of proteolytic enzymes to avoid some of the disadvantages inherent in the aging process. When premortem injection is used, the animal's circulatory system distributes the tenderizing enzymes. This has some drawbacks in that the enzyme sometimes becomes clogged in tiny capillaries and also has a tendency to accumulate in areas which have a rich blood supply, such as the liver and kidneys. To avoid these problems, meat scientists rapidly inject the enzyme, and the animal is promptly slaughtered. Best results are seen when the animal is killed after a time no longer than 60% of the time needed for one cycle of the animal's circulatory system. In cattle, this is just under a minute; in large roosters and turkeys, it is 2 to 3 seconds.

If the tenderizing enzyme is introduced after slaughter, it should be done before the onset of rigor mortis, when the carcass is still warm, to ensure distribution from the point of injection. About 3% (by weight) of water injected into the carcass aids the enzyme in providing uniform tenderization.

Microorganisms can be used to enhance the flavor of meat without tenderizing it. If water is injected along with the organism, tenderization also results.

Enzyme-Catalyzed Sugar Conversions

A second important class of nutrients is the carbohydrates—starches and sugars. The enzyme amyloglucosidase (also called glucoamylase), obtained from Aspergillus mold, is used in most commercial processes for the conversion of starch to glucose—an easily metabolized sugar. (Figure 5–1 illustrates starch structures; Figure 6–1 illustrates glucose structures.) In a typical commercial process, cornstarch is thinned by heating a water slurry to temperatures above the point at which the starch gelatinizes (becomes gellike). Dilute acid or the enzyme alpha amylase can then be used to partially hydrolyze the starch, forming materials called dextrins. The solution pH is adjusted to 4.0 and glucoamylase is added. The dextrin solution is cooked with the enzyme at 55° C to 60° C for 24 to 96 hours depending on the amount of enzyme used. Glucoamylase is essentially deactivated at the end of the incubation. After the conversion of dextrins to glucose is completed, the solution is filtered and treated with resins to remove impurities. Some water is removed by evaporation and crystalline glucose is added. This acts as a seed which induces crystallization of the glucose in solution.

Various research groups around the world have studied methods for immobilizing amyloglucosidase to permit the continuous conversion of starch to sugar. The worldwide replacement of the soluble enzyme process, by one employing immobilized amyloglucosidase, appears imminent.

Pure starch is made up entirely of glucose molecules held together by chemical bonds. There are two basic forms of starch, as shown in Figure 5–1, a linear form and a branched form. All of the glucose molecules of starch, except one on an end, have carbon 1 tied up in bonding to neighboring glucose molecules. The single free carbon 1 in each starch molecule is the carbon which can be shown as an aldehyde (CHO) group. This end is therefore called the "reducing end." Because

Figure 6–1   Alternative forms of glucose structures.

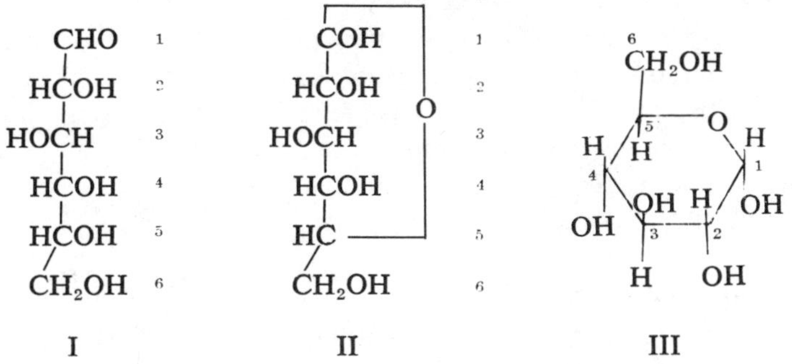

Figure 6–1 show diagrams used by chemists to describe the glucose molecule. Diagram III is used most frequently, but I is used to explain the reducing activity of glucose. The sugar is able to reduce an ammonia solution of silver nitrate, forming free silver. This reducing activity is due to the CHO (aldehyde) group shown at carbon 1 in diagram I.

Two glucose molecules bonded together as shown make a molecule called maltose:

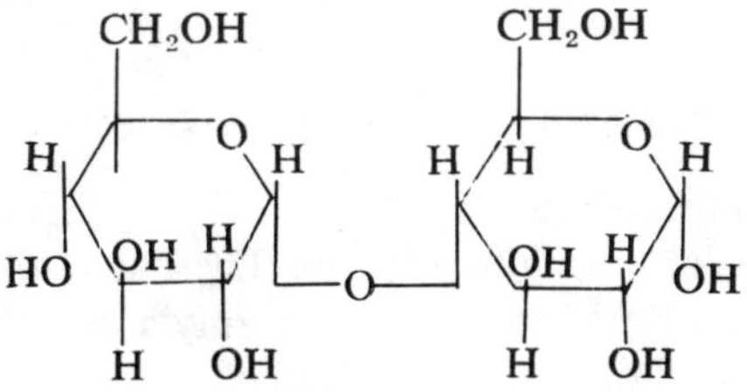

of its branched structure, amylopectin has many nonreducing ends, but it has only one reducing end.

Starches can be converted to syrups by use of enzymes known as amylases. The most important amylases are alpha amylase, beta amylase, and glucoamylase (amyloglucosidase). Alpha amylase attacks starch randomly at 1,4–bonds, producing much smaller molecules called dextrins. Beta amylase attacks only at nonreducing ends, splitting off two glucose units at a time. The two-glucose sugar produced is called maltose. Its structure is illustrated in Figure 6–1. Neither alpha nor beta amylase can attack 1,6–bonds but alpha can bypass them, attacking any 1,4–bond in the molecule. In doing this, it generates many nonreducing ends for beta attack. Glucoamylase attacks 1,4–bonds preferentially, but it attacks 1,6–bonds at a reduced rate.

Various proportions and combinations of the three amylases are used together or successively to prepare syrups having properties desirable for use in different industries. Acid hydrolysis is sometimes used for the early phases of syrup production, just as it is in the commercial glucose-making process. Syrup production from starch represents an intermediate stage between unconverted starch and complete conversion to glucose.

Chocolate syrups are made by decomposition of cocoa starch by a mixture of amylases of fungal origin. A slurry of cocoa in water is adjusted to the proper level of acidity (pH 6.2 to 6.6). It is heated to 185° F to 190° F and kept at that temperature for 30 minutes. The temperature is raised to 200° F and held for a few minutes. The mixture is then allowed to cool. Dry sucrose or sucrose syrup may be added to hasten the cooling. When the temperature gets down to 140° F, a solution of the fungal amylase is added. During a reaction period of about 30 minutes, the enzyme decomposes the cocoa starch to a mixture of dextrins, maltose, and glucose. Chocolate syrups produced by enzymes have particularly good stability properties, good flavor, and are easily dispersible in cold water.

Enzymes in the Baking Industry

During milling, about 3% to 4% of the starch granules of hard wheat are damaged in normal operation. These damaged granules represent vulnerable points for attack by the amylases, which are the most important enzymes in the baking industry.

Normal flours have enough beta amylase but are deficient in alpha amylase activity. The combined action of the two enzymes is important in forming sugar for fermentation by the yeast. The addition of small amounts of alpha amylase to bread dough produces normal, medium-strength flour dough and results in better crust color, grain, and volume of bread loaves. Larger amounts of alpha amylase increase the volume of the loaf, but the grain and texture are inferior. The enzyme attack on the damaged granules starts at the time of mixing and continues until the enzyme is denatured during baking.

Fermentation is an essential part of the breadmaking process. It is in fermentation that the dough becomes workable and develops the desired qualities of grain, texture, loaf volume, and flavor.

There are two basic methods for bread making—the straight

dough method and the sponge dough method. In the former, all ingredients—flour, sugar, shortening, dry milk, salt, yeast, water, and any added alpha amylase—are mixed into a pliable dough before fermentation. In the sponge dough method, about three-fourths of the flour is mixed with yeast and water to form a stiff dough known as a "sponge." The sponge is fermented for a relatively long time and is then mixed with the rest of the flour and the other ingredients. A "rest period" of 15 to 30 minutes follows this mixing. Sponge dough produces better processability and improved grain and tenderness. Timing of steps is essential once the sponge has been mixed with the remaining ingredients.

The added alpha amylase produces a number of advantages including decreased mixing time, improved gas retention, acceleration of development of dough extensibility and processability, and reduced proof (rising) time and pan flow.

Adequate gas production in fermenting dough is of paramount importance to the quality of the finished bread. Gas production depends on the amount of maltose (fermentable sugar) available. This maltose is made by the action of the natural amylases of the flour with the assistance of the supplementary added alpha amylase.

Fungal proteases are sometimes added to sponge doughs. These act on the gluten and reduce dough mixing time by about one-third. (It is believed that the most frequent defects in homemade bread result from insufficient gluten development due to inadequate mixing.) Protease-added doughs are more extensible, rise faster because of better gas retention, and produce more symmetrical loaves with improved grain, texture, and flavor.

After fermentation (and resting of sponge dough), the dough is divided, rounded, and allowed to rise before being molded into designated shapes. The shaped doughs are allowed to rise in the baking pans for 50 to 70 minutes at 100° F to 105° F and 90% to 95% humidity. They are then baked at 375° F to 400° F for 18 to 30 minutes. During the baking process, the enzyme activity increases to a maximum and goes rapidly to zero as the heat denatures the proteins.

GLUCOSE ISOMERASE

Glucose isomerase is another enzyme which has perked worldwide interest and research. This enzyme converts glucose to fructose—a sugar which provides twice as much sweetening power as glucose and no more calories. Equation 6–3 illustrates the conversion.

The major sources of glucose isomerase are microorganisms of

Equation 6–3

$$\text{D-glucose} \quad \xrightarrow{\text{Glucose isomerase}} \quad \text{D-fructose}$$

the streptomyces family. Freeze-dried streptomyces cells have been used for the commercial production of high-fructose syrups using high-glucose syrups as substrates.

The current interest in immobilized enzyme technology and the commercial importance of the glucose isomerase reaction made immobilized glucose isomerase a highly desirable commodity. Many publications and patents describe glucose isomerase immobilized by various methods on a variety of supports.

Commercial use of the immobilized enzyme in the United States is licensed to an American manufacturer by the Japanese company which developed the process and patented it.

Invertase in the Candy Industry

Sucrose, the common table sugar derived from sugar cane or sugar beets, can be converted to a mixture of glucose and fructose by an enzyme known as invertase. This reaction is illustrated in Equation 6–4.

A patent issued in 1932 to Wallerstein Laboratories described an interesting process for candy making that is still in use today. An enzyme is used in the manufacture of cream-centered candies. Small amounts of invertase are added to fondants containing enough sucrose to give the firmness required for machine casting. During storage, the invertase converts the sucrose to glucose and fructose—a mixture frequently called "invert sugar." The process reduces the moisture content in the cream of the confection. As the chemical equation (6–4) shows, a molecule of water is used for each molecule of sucrose converted. The content of the soluble solids is also increased, in that one part of glucose and one of fructose are formed for each part of sucrose converted. The soluble solids increase to the point where the center does not dry out and

Equation 6–4

$$H_2O \text{ and Water and Sucrose} \xrightarrow{\text{Invertase}} \text{D-glucose and D-fructose}$$

the moisture content is low enough to prevent yeast growth. Fermentation in chocolate-coated confections usually occurs only when the content of solids is less than 79%. Wallerstein's purified invertase, when used under proper conditions (1 ounce of enzyme per 100–pound batch of candy), produces 82% to 83% solids. The humectant (moisture-holding) properties of fructose assure that the product will stay moist. Sucrose has no humectant properties; if sucrose is used alone, the confection will lose last traces of moisture and will appear stale.

The enzyme is used in larger proportions in making products such as cherry cordial creams. The fruit is embedded with a suitable solid sugar mixture and invertase is added (4 ounces per 100 pounds). It is kept cool and coated with chocolate. The confections are stored at slightly elevated temperatures causing the enzyme to slowly liquefy the interior. It is essential that the solids' content of these candies be raised as quickly as possible to prevent fermentation, which would produce gas and burst the product.

Enzymes in the Brewing Industry

The brewing process can be broken down into a number of discrete stages: malting, mashing, fermenting, lagering, and bottling or racking.

The malting process consists of controlled germination of raw barley followed by drying. During germination, enzyme levels increase substantially. Proteases are formed and they attack other barley proteins during germination. Hemicellulases develop and split cell tissues making starch available for attack by amylases. High levels of alpha amylase arise during germination and beta amylase increases somewhat. The amylases are the most important enzymes in brewing. They break down starch molecules forming sugar, which is converted to alcohol and carbon dioxide by fermentation. The activity of the amylases is greatest during mashing—the central process of brewing.

In this process, the malt and some adjunct grain, such as corn or rice, are crushed and mixed with water. Crushing makes material available for solution and for enzyme attack. The brewing mixture at this stage is called "mash." Approximately 80% of the dry matter of malt is extracted into the water giving a solution called "wort."

After separating the wort from the insoluble materials (spent grain), the solution goes to a wort kettle where it is vigorously boiled with hops for 1½ to 2 hours. Hops are not only responsible for the aroma and characteristic bitter taste of beer, but they also increase the biological stability and improve retention of the foam head.

The wort is cooled and filtered to remove all sludge. Yeast is added and fermentation is carried out at 6° C to 10° C for 8 to 10 days. Cooling is necessary since fermentation evolves heat. If the temperature goes above 11° C to 12° C, taste and head retention of beer may suffer and bacterial contamination becomes more likely.

After this "primary fermentation," the product is called beer, but does not taste like it. A secondary fermentation occurs during lagering (storage) at low temperatures in closed vats or tanks for 1 to 2 weeks. This secondary fermentation starts vigorously and evolved carbon dioxide carries off some volatile alcohols and aldehydes. This reduces the "young" bouquet of "green" beer. Remaining alcohols and acids are converted to esters by yeast esterases. This contributes to the taste and aroma of the product. The secondary fermentation contributes carbonation (carbon dioxide saturation) to the beer.

After lagering, the beer is filtered, and "chillproofing" agents are often added. These are proteolytic enzymes, such as papain and pepsin, which decompose protein-tannin compounds suspended in the beer. If not removed by the addition of enzymes, these suspended materials produce a turbidity known as "chill-haze," which is unacceptable to the American beer drinker.

The beer is bottled or racked (put into a keg) under pressure with carbon dioxide. Bottled and canned beer is pasteurized in the container; draught beer is usually not pasteurized.

The brewing process just described is the classic process. Modern enzymic methods make it possible to improve the process and to use cheaper materials. Much of the barley malt can be replaced by cheaper adjunct grains such as corn, rice, or unmalted barley. Proteases must be added to replace those which would have been formed in malting. Fungal amylases are also added to decompose the starches of the adjunct grains and provide more sugar for fermentation. Beers produced by this method are comparable in analytical and taste properties to beers produced by the classic process.

PECTIC ENZYMES IN FRUIT JUICES

All fruits contain pectin, a high molecular weight carbohydrate. It acts as a protective agent, stabilizing extremely small pulp particles in the juice and causing them to remain suspended, resulting in a cloudy mixture. Attempts at filtering pectin-containing juices invariably result in clogged filters and little or no juice clarification. Pectic substances confer "body" to juices.

Although fresh juices naturally contain some pectic enzymes, the amount is so low that spoilage or fermentation usually occurs before natural clarification. The addition of more pectic enzymes accelerates the clearing process. Pectic enzymes added to juice decompose the pectin, and the pulp particles coalesce and settle out. The juice can then be filtered very easily, giving a clear, sparkling product. The enzymes have no deleterious effects on the delicate elements of flavor and aroma. After their job is finished, the enzymes can be inactivated by heating the juice to 170° F for 5 minutes.

Juices from apples and citrus fruits are often used without any clarification. Pectic enzymes can be used, though, if a brilliantly clear juice is desired. Soft fruits and berries cannot easily be pressed for juice before very thorough enzymic pectin removal, however.

As illustrated in Figure 6–2, the most important pectic enzymes are pectin esterase and polygalacturonase. The former attacks the methyl ester, while the latter acts on bonds holding the individual sugar units together.

PECTIC ENZYMES IN WINE MAKING

The preparatory steps to wine making quickly follow the grape picking. Stems are removed and grapes are crushed. If a white wine is desired, skins are separated from the juice immediately after crushing.

Grape juice, called "must" in the wine industry, serves as a source

Figure 6–2   Pectin.

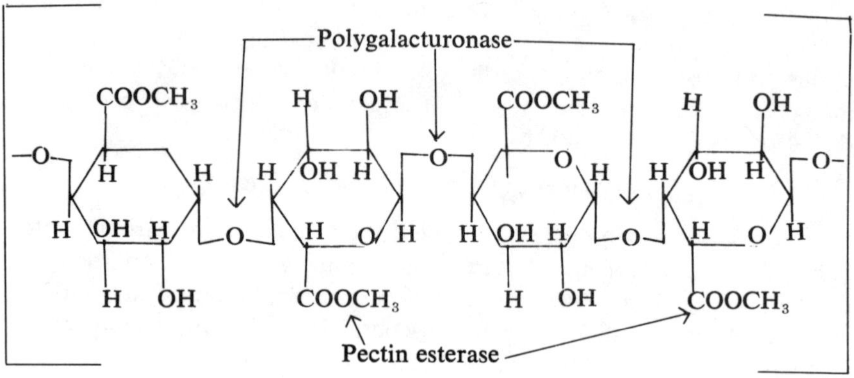

The pectin structure is shown in brackets to indicate that only a portion of the molecule is illustrated. The portion shown, repeats itself many times in a linear chain of identical units. The principal enzymes attacking pectin are polygalacturonase and pectin esterase. The positions attacked by these enzymes are indicated.

of sugar, which is fermented by added yeast, forming alcohol. Sometimes it is necessary to add extra sugar to permit fermentation to generate an alcohol concentration of 9%—the minimum concentration needed to prevent rapid spoilage.

The must is sometimes allowed to settle for 24 to 36 hours at 50° F before fermentation. Skins and other solids settle to the bottom and the clear must is taken off to ferment. The must is inoculated with a pure yeast culture. For a dry, white table wine having less than 14% alcohol, fermentation is carried out in closed vessels at temperatures no greater than 60° F. The high acidity of grape juice (pH 3.0 to 3.6) prevents the growth of spoilage bacteria, while permitting free growth of fermenting yeasts. Sugar content is measured daily. As the fermentation nears completion, wine is drawn off (racked) into another container which should be filled completely. Every day until fermentation ceases, gas pressure should be released. Wine will settle in approximately 6 weeks at 50° F. (Yeast cells and other insoluble materials settle to the bottom.) Wine must be racked from this sediment within 6 weeks or the spontaneous breakdown of yeast cells will cause liberation of noxious hydrogen sulfide and other undesirable contaminants.

The wine is rough-filtered early in the year. Later, during the spring, it is chilled to 25° F and held at that temperature for a few days to precipitate excess tartrates. It is racked and clarified with a slurry of bentonite clay, which adsorbs impurities. The wine may then be filtered directly into the bottle.

Pectic enzymes are used advantageously in wine making. Addition of such enzymes to crushed grapes increases the volume of free-flowing juice, reduces pressing time, and increases total juice yield. The enzymes also increase the extractability of coloring agents from the skins, thus improving the color of red wines. By decomposing the pectins, they hasten filtration; less bentonite is needed. Aging is accelerated and no sediment forms on storage. Stability of the shelf life is increased. The only disadvantage is a slight elevation of the methyl alcohol content from approximately 0.019% in untreated wine to 0.023% in wines treated with pectic enzymes—both values well below permissible levels.

GLUCOSE OXIDASE IN THE FOOD INDUSTRY

Glucose oxidase is useful in the food industry because of its ability to remove glucose or oxygen from a system. Equation 6–5 illustrates

Equation 6–5

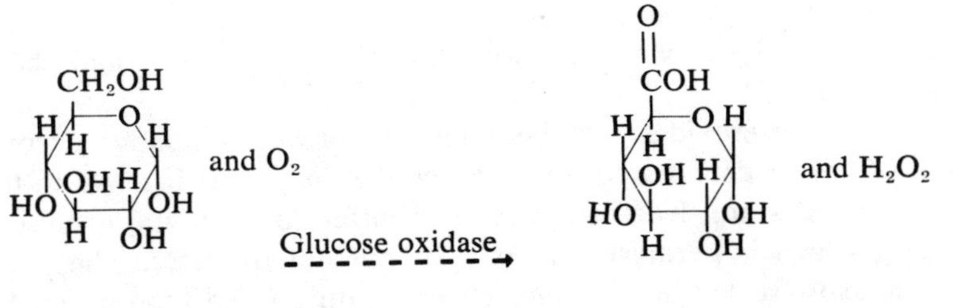

D-glucose and Oxygen $\xrightarrow{\text{Glucose oxidase}}$ Gluconic acid and Hydrogen peroxide

the reaction. The commercial enzyme, obtained from a mold, Aspergillus niger, contains some catalase activity. This is generally beneficial since the catalase decomposes hydrogen peroxide. Extra catalase is often added.

Glucose oxidase is used to remove glucose from egg whites and whole eggs before drying. Powdered egg products deteriorate during storage because of a "browning reaction" between glucose and proteins. Such deterioration causes loss of whipping properties, proteins precipitate, and off-flavors develop. The best method for stabilizing dried egg products is the removal of glucose before drying. Glucose oxidase does this most satisfactorily.

Oxygen is a reactive gas which causes many types of food deterioration, including flavor and color changes. Glucose oxidase has been found effective in stabilizing foods and beverages by removing oxygen from the food or container.

Breweries have tested glucose oxidase for eliminating traces of oxygen from beer. Tests have been consistently successful and taste panels preferred the flavor of the enzyme-treated beer to that of untreated brews.

Mayonnaise is the classic example of a food item which can be protected from deterioration by glucose oxidase. This product is an emulsion of oil in water containing much air, packed in a clear glass container, and exposed to conditions ideal for causing deterioration and rancidity. The addition of glucose oxidase (50 parts per million) gave a stabilized product which showed no detectable color change, rancidity, or increase in peroxide during 6 months storage. Untreated samples stored in the same manner had faded and were rancid as early as the third month. After being stored for a year, enzyme-treated samples were still judged acceptable for use while the untreated samples were completely spoiled—very rancid, with the emulsion broken and the color badly faded.

Glucose oxidase has been used to prevent oxidative decomposition of dry or dehydrated foods. Since the enzyme is inactive in the absence of moisture, it is dissolved in a buffer solution and enclosed in a packet which is permeable to oxygen but not to water. The packet is hermetically sealed in a container along with a dried food product, such as whole milk powder, cake mixes, and nut meats. Such packets completely remove oxygen within 24 hours and they substantially increase the storage stability of the products.

A Precooked Dry Cereal Process Using Enzymes

A process for preparing a cereal product having a custard-like consistency has been described in a patent issued to Salada-Shirriff-Horsey, Inc. Selected cereal flours are mixed with water and heated to gelatinize the starches. An amylase enzyme is added to convert some of the starch to sugar and dextrins. The enzyme reaction is stopped after 10 to 20 minutes by heating to denature the enzyme. Mineral salts and vitamins are added and the product is cooked and dried. This heating further gelatinizes the starches so that they will more readily absorb water. The dried cereal is broken into discrete particles which are mixed with a rennet-containing preparation before packaging. The rennet preparation contains sugar, gum arabic, water, and a calcium salt in addition to the enzyme rennet.

In use, 4 level teaspoons of the cereal product are mixed with ½ cup of milk at 110° F. After 1 minute of stirring, it is allowed to stand for a minute or longer to give the enzyme time to "set" the milk. The custard-like cereal has a light, creamy apearance. It does not dry out or become pasty on standing and it remains homogeneous without constant stirring.

Enzymes as Digestive Aids

Enzymes are important and useful aids in the preparation and processing of foods, but they are also of critical importance in digesting the food we eat. Digestive enzyme levels decrease as we get older and, in some cases, after surgery. Fortunately, there are available commercial preparations of digestive enzymes to alleviate these deficiencies. They will be discussed in Chapter 10.

Addendum: A process for the immobilization of glucose isomerase has been developed and patented by Messing and colleagues at Corning Glass Works. This technology has been licensed to CPC International, Inc., and has been used industrially by them since 1976.

Chapter Seven

Enzyme Uses in the Dairy Industry

MILK IS A complex solution. A typical analysis of cow's milk shows 3.8% fat, 3.2% protein, 4.8% carbohydrate, 0.7% minerals, and 87.5% water. The fat occurs in globules containing thousands of different molecules. The globules are enclosed by a complex membrane acquired at the time the milk is secreted. The most abundant proteins of milk are casein, albumins, and globulins. Casein, a phosphorus-containing protein, comprises about 80% of the milk protein. It forms the "curd" in cheese making. The albumins and globulins are the whey proteins; they remain in solution when the curd separates.

ENZYMES IN MILK

Although they do not occur in appreciable concentrations, at least 20 enzymes have been found in milk. Most of them have little, if any, effect on the milk. Some are thought to "spill over" from the blood, while a few are probably formed in the udder for a particular use—possibly as digestive aids for the suckling young.

Lactalbumin is a milk protein that is necessary for optimal activity of lactase synthetase—the enzyme which makes lactose (milk sugar). In the absence of lactalbumin, lactose synthetase forms molecules called galactoproteins.

Some of the enzymes of milk protect it against microorganisms. Lactoperoxidase is a powerful inhibitor of microbial growth when it is provided with a supply of hydrogen peroxide. The action of xanthine oxidase in the milk generates this peroxide so the two enzymes work together as a defensive system. An antimicrobial enzyme called lysozyme is found in low levels in cow's milk and in higher levels in human milk.

Cold Pasteurization

Pasteurization is an effective technique for destroying pathogenic organisms in milk, but it has some disadvantages in treating milk intended for cheese making. The process destroys acid-forming bacteria and inactivates some enzymes, such as lipases, proteases, and phosphatases, which are desirable in the cheese-making industry. Hydrogen peroxide has been used to destroy milk pathogens. It effectively reduces the pathogen count while permitting the lactic acid forming microorganisms to survive. (Lactic acid formation is a preliminary step in cheese making.) The desirable enzymes are not affected by the dilute solutions of hydrogen peroxide. Residual hydrogen peroxide would interfere with the growth of "starter organisms" used in cheese making, so it must be removed.

The enzyme catalase decomposes hydrogen peroxide forming oxygen and water. This process using hydrogen peroxide is called "cold pasteurization"; it is permitted in the United States for treating milk used in cheese making. Two methods are used. In the "flash" method, about 0.03% of 35% hydrogen peroxide is added to cool milk, which is then passed through a pasteurizer, where it is heated to 52° C for 25 seconds. It is then cooled to the setting temperature of the cheese and excess catalase is added to destroy remaining peroxide. In the "vat" or "batch" method, 0.06% of 35% hydrogen peroxide is added to the vat of milk at the normal setting temperature (31° C to 32° C). After 20 minutes, catalase is added to destroy the peroxide.

Tests are run to be sure that there is no residual hydrogen peroxide in the milk. When these tests are negative, it is safe to add the starter cultures for cheese making.

Cheese Making

Like alcoholic fermentation, the art of cheese making has been practiced for thousands of years. Both are mentioned in the Bible. Legend has it that cheese was discovered by accident. An Arabian merchant, carrying milk in a pouch made of a sheep's stomach, traveled across the desert on his camel. The enzyme, rennet, found in the lining of the sheep's stomach, the hot desert sun, and the agitation produced by the camel's gait all combined to transform the milk into a mixture of curd and whey. The "inventor" satisfied his thirst with the nutritious whey and his hunger with the curd. Cheese was born.

For centuries, rennet from the stomachs of calves, lambs, and kids had been used for cheese making. Young animals were used because if the animals are fed anything but milk, pepsin is found in the stomach along with the rennet. Until the recent decrease in availability of rennet, pepsin was considered undesirable in the cheese-making process. Although pepsin and other proteolytic enzymes are capable of separating milk into curd and whey, an undesirable bitterness develops upon storage when proteolytic enzymes other than rennin are used. To see why this may happen, it is instructive to look at the nonenzyme proteins of milk and the mode of action of the proteolytic enzymes on these proteins.

As we have seen, the most abundant milk proteins are casein, albumin, and globulin. Of these, only casein is coagulated by proteolytic enzymes to form curd. The albumin and globulin remain in solution as the whey proteins.

Casein is a phosphorus-containing protein which exists in four distinct types designated by letters of the Greek alphabet—alpha, beta, gamma, and kappa. In milk, these casein molecules aggregate into clusters called micelles. Each subunit of the micelle probably contains each of the four caseins. In the absence of kappa casein, alpha, beta, and gamma caseins can be precipitated by calcium ion. Kappa casein contains a highly soluble carbohydrate portion known as sialic acid, which prevents precipitation by calcium ion. In milk, kappa casein acts as a protective agent preventing precipitation of the casein micelles.

Rennet seems to attack kappa casein specifically, splitting off the sialic acid along with a chain of amino acids (a polypeptide). The part of the kappa casein remaining is called para-kappa casein. The removal of the sialic acid is enough to destabilize the soluble casein micelle and curd formation results. The proteolytic action of rennin seems to stop at this point. Other proteolytic enzymes are able to split the kappa casein, but their action continues producing more degradation of the milk proteins. It is this increased degradation which causes the bitter flavor observed when some proteases replace rennin in cheese making.

Acceptable Rennin Substitutes

Historically, male calves have been used as a source of rennin (rennet) in the United States. Females were grown for milk production.

The number of calves slaughtered has gone down drastically in recent years because of increased consumer demand for beef and because of improved methods for raising calves for beef at reasonable cost. This has happened at a time when cheese production is increasing rapidly. Thus, it became necessary to find a replacement for rennet.

Pepsin, the primary curd-forming enzyme isolated from the stomachs of mature cattle, has a number of disadvantages. It takes 10 times as much pepsin as rennin to form the same amount of curd and the product tends to develop a bitter taste on storage. Nevertheless, it is used, along with other proteolytic enzymes, in curd-forming blends. Trypsin is capable of coagulating milk but its subsequent proteolytic action decomposes the curd. A number of bacterial proteases have been found to have rennin-like activity. Typical of these is "Emporase"—a microbial rennet derived by pure culture fermentation by a mold called "Mucor pusillus Lindt." This enzyme has been tested and proven in the manufacture of all major varieties of cheese. It can be used interchangeably with veal rennet or pepsin and can be blended with either of them. Most cheese made in the United States today is made using enzyme blends such as rennet-pepsin, rennet-"Emporase" or "Emporase-pepsin." "Metroclot" is a commercial pepsin recommended for use in conjunction with rennet.

Some studies of immobilized rennet and immobilized trypsin have been conducted, but results have been less than spectacular. Problems observed include leaching of the enzyme from the support, relatively rapid inactivation of the enzyme during use, and plugging of enzyme-containing columns. None of these factors is insurmountable, but they represent obstacles which must be overcome for successful commercial use of immobilized coagulating enzymes.

Curd Formation and Processing

After the milk has been pasteurized (or "cold pasteurized"), and before the addition of rennin, acid is developed in the milk by use of a "starter"—a bacterial culture which converts lactose to lactic acid. A period of 1 hour at 72° F is allowed for the acidity to develop. The coagulating enzyme (rennin or a rennin substitute) is then added to cleave the casein and form curd. The firmness of the curd can be controlled by regulating the kind and amount of enzyme used, the degree of acidity, the temperature, and the reaction time. Whey, the solution remaining after the curd separates, is usually drained off. The curd is cut into small pieces, pressed, and heated to remove more whey. It is usually salted before the ripening process is started. The salt seasons the cheese, retards the growth of bacteria, helps in separating curd from whey, and helps regulate the ripening process.

Ripening the Curd

The treated curd is now ready for ripening—the process of becoming cheese. During the ripening (or aging) period, enzymic reactions produce the characteristic cheese flavors. Proteolytic activity is particularly great due to starter bacteria and the coagulating enzyme. Among the changes occurring during ripening are: (1) multiplication of bacteria and changes in the types of bacteria; (2) decreases in the sugar, acidity, and moisture contents; (3) changes in flavor and body; and (4) gas production in most varieties. Ripening is usually carried out at 40° F to 60° F and a relative humidity of 70% to 75%. The time of ripening varies from none (for cottage cheese and similar varieties) to months or years.

Lipases in Italian Cheese Making

The characteristic piquant flavor of Italian cheeses has been attributed to the action of enzymes called pregastric esterases. The cheeses have always been made using a rennin paste which contains some of the pregastric enzymes as a contaminant. Modern technology has made available commercial supplies of these enzymes, which are extracted from oral glands of calves, lambs, and kids. Pregastric esterases are now used along with coagulating enzymes in making Italian cheeses. They act as lipases—enzymes which decompose fat molecules. In the cheese-making industry, they act on butterfat, producing a specific, reproducible ratio of free fatty acids which cannot be duplicated by other enzymes having lipase activity. The availability of the pregastric enzymes thus makes it possible to continue producing Italian-type cheeses despite the shortage of rennin.

THE ACID WHEY PROBLEM

As cheese production continues to increase, there is a corresponding increase in the production of whey; 10 pounds of milk yields about 1 pound of cheese and 9 of whey. Modern Americans, unlike the legendary Arabian traveller, do not drink whey. This is unfortunate for a number of reasons. Whey is a highly nutritious material which contains half the solids of whole milk. It is an abundant, protein-rich foodstuff in a protein-starved world. Yet, more than half of the whey produced is wasted. In a sense, it is worse than wasted, for it contributes a highly significant pollution problem. Whey is the largest by-product of the dairy industry and is said to be one of the most troublesome by-products of any industry. The waste of whey is largely due to the high cost of processing the water solution to isolate the desirable components. This high cost is reflected in rising cheese prices.

Approximately 57% of the cheese produced in the United States is processed cheese—mostly cheddar. This is made from whole milk and the resulting whey—called sweet whey—has a pH of 5 to 7. Most of the remaining 43% is cottage cheese, which is prepared from skim milk. It yields "acid whey" with a pH of 4 to 5. This is a less desirable product and harder to handle than sweet whey. It is estimated that 43¼ million pounds of recoverable protein are wasted annually in cottage cheese whey. The estimated value of this protein is $1.50 per pound.

Whey is a greenish yellow fluid. It contains about half of the solids of whole milk from which it is derived and most of the water-soluble vitamins and minerals. Most of whey is water. A typical analysis of the remaining 6.3% reveals that 68% to 72% is lactose (milk sugar), 12% to 13% is protein, 8% to 9% is minerals, and there are small amounts of vitamins, fats, and lactic acid.

The whey, which is not used, presents a serious disposal problem. It is a costly pollutant in the nation's waterways, and some communities have imposed a surcharge on dairies disposing of whey in sewage. In terms of sanitary engineers and other pollution specialists, whey has a high "biological oxygen demand—BOD." Every thousand gallons of whey discharged into a stream requires for oxidation the dissolved oxygen in 4.5 million gallons of unpolluted water. This oxygen is needed to sustain aquatic life, kill harmful bacteria, and purify water.

Whey disposal would not be a problem if it were easy to isolate whey nutrients in a usable form, but it is not. The nutrients exist as soluble components in a solution that is mostly water. Modern physical techniques of separation called ultrafiltration and reverse osmosis have recently been introduced into dairies. These techniques are useful for recovering the valuable milk protein. The waste product remaining, to be used or dumped, is a solution of lactose containing small quantities of lactic acid and vitamins, such as riboflavin.

Lactose Problems

Lactose is of little commercial value. It would be desirable to recycle it through the dairy for use in ice cream, eggnog, yogurt, and other dairy products needing solids and sweeteners. This is not practical for three reasons: lactose is not very soluble; it is not very sweet; and most people cannot digest it.

Because lactose is not very soluble, high concentrations of it result in the formation of large crystals, which cause a sandy or gritty texture in such products as condensed milk, ice cream, or ice milk. These crystals also tend to settle out, forming a deposit at the bottom of containers.

If sucrose is assigned a sweetening power of 100, the relative sweetening power of lactose is 16. Yet, lactose has as many calories as an equal weight of sucrose. Thus, lactose cannot be considered a desirable sweetening agent.

The enzyme responsible for the digestion of lactose is an intestinal protein, commonly called lactase. Adults of most animal species lack this enzyme. Most humans after age 2 to 4 are also deficient in this catalyst, which is essential for lactose digestion. In general, Northern Europeans and white Americans do not suffer from this deficiency, but they are the exception. Ingestion of lactose produces several pathological symptoms in the lactase-deficient individual. These include bloating, gas, abdominal cramps, and severe diarrhea with frothy stools containing lactose and lactic acid. Lactase deficiency in infants is rare, but it produces severe consequences. After ingesting milk (from mother or bottle), the lactase-deficient infant suffers watery diarrhea, which can lead to dehydration and severe malnutrition. Continued milk feeding can cause death.

It can be seen that the use or disposal of whey is not the only problem involved here. In our protein-starved world, dry milk represents a potentially useful source of high-quality protein. Yet, it is the very people who need this protein most desperately (the people of the "developing nations" and the "backward nations") who are least likely to be able to use it. The fact of protein malnutrition complicates the picture further, for the condition increases the likelihood that its victims will not be able to digest lactose. Although the protein of dry milk would help these starving people, the lactose would poison them.

The Light at the End of the Tunnel

The enzyme beta galactosidase (more commonly called lactase) breaks down lactose, forming glucose and galactose, as illustrated in Equation 7–1. The isolation of this enzyme from microorganisms makes it available in large quantities at a reasonable price. Lactase has been

Equation 7–1

and H$_2$O   Lactase ⟶   and

Lactose and water   Glucose and Galactose

used to some extent for the preparation of low-lactose products. Such products are indistinguishable from untreated materials except for a somewhat sweeter taste.

In recent years, because of dramatic price increases for sweeteners, the cost of processing whey has been outweighed by the potential economic gain. A large, highly funded research effort has been aimed at developing immobilized lactase systems for the treatment of whey. This research has been carried out by industrial firms, such as Corning Glass Works; at universities including Cornell, Lehigh, Rutgers, and Wisconsin, among others; and at government laboratories, such as the United States Department of Agriculture laboratories.

A recent United States patent issued to Weetall and Yaverbaum and assigned to Corning Glass Works describes the treatment of whey with immobilized lactase and immobilized glucose isomerase (see Chapter 6). This treatment removes most of the lactose, converting it to glucose and fructose, sugars of greater sweetness, solubility, and digestibility.

The enzymes are separately bonded to porous carriers having a high surface area such as zirconia-treated glass particles. The glass-supported enzymes are packed into columns for treating the waste materials. Whey may be filtered and added directly to the lactase column to decompose the lactose, but a preferred procedure involves treating the whey by ultrafiltration and reverse osmosis to first recover the milk protein. Most of the salts and vitamins are also removed by physical methods, leaving a water solution of lactose and lactic acid. This solution is passed through the column containing glass-supported-lactase. Most of the lactose is converted to glucose and galactose—a mixture of sugars that can be used directly in the dairy or elsewhere. A preferred option is to raise the pH to 7.5, add a magnesium salt, and pass the solution through the column containing glass-supported glucose isomerase. Additional glucose may be added to the solution before it traverses this column. The glucose isomerase converts glucose to fructose, which is twice as sweet. The final product from this sequence is a solution containing fructose, glucose, galactose, and a little lactose. This may be used as a sweetener for ice cream or other consumable products.

A fringe benefit has been realized in the study of lactose decomposition by immobilized lactase. Cheese made from milk, in which the lactose has been hydrolyzed, ages much faster than cheese made from nontreated milk. The presence of lactose may inhibit the aging process.

Approaches similar to that described in the Corning patent have been taken by other research groups. The principal differences involve the type of support used, the method of attachment of the enzyme(s) to that support, and the type of reactor used. While at Lehigh University, I worked on a project supported by the National Science Foundation aimed at solving the whey and lactose problems. The first published account of an enzyme bonded to stainless steel was one result of this work.

Lactases derived from bacteria, yeasts, and fungi are commercially available. Although all of these enzymes catalyze the hydrolysis of lactose, other properties differ. For instance, there are differences in pH and temperature optima. These differences make it possible to select an enzyme having properties desirable for a specific application. For instance, if the treatment of acid whey is the goal, an enzyme having an optimum pH in the acidic range is desirable; sweet whey would preferably be treated with an enzyme having most activity at a higher pH.

While the development of lactose-conversion processes will primarily benefit the dairy industry, other food industries will also be helped by it. A new and abundant source of effective sweeteners is made available for syrups and other sugar applications. The bread-making industry can also be served. The skim milk used in this industry contains lactose, which is not fermentable by yeast and does not provide sweetness. By lactase treatment, the lactose is converted to glucose, which is readily fermentable by yeast, and galactose, which aids in the development of crust color in bread.

Hazards of This Technology

It would be inappropriate to describe the benefits of the developing lactase technology and to ignore the hazards. The one important caveat involves the ingestion of high levels of galactose—one of the products of the lactase reaction. This sugar has been indicted as a causative agent in the development of cataracts and in conditions of impaired growth. In addition, there is a genetic disease known as galactosemia, which is characterized by an inability to metabolize galactose. Adverse effects are seen in the eyes and liver and there is a high incidence of mental retardation in victims of this congenital disease. These people should not use dairy products—treated or untreated. Treated products provide galactose directly; untreated products provide lactose, which acts as a poison if not metabolized and which provides galactose if it is metabolized.

Yet, the development of lactase technology must be considered a positive achievement. We may see the time when all dairy products are routinely treated for conversion of lactose. Future generations of scientists may busy themselves with projects aimed at removing galactose from dairy products.

OXIDATIVE FLAVOR OF MILK

A problem of long standing in the dairy industry is the prevention or removal of "oxidized flavors" in milk. This flavor develops spontaneously and presumably arises from the action of oxidative enzymes in the milk.

Proteolytic enzymes have been found capable of retarding or preventing the development of the oxidized flavor. The use of such enzymes has been limited, however, because of high cost and because of difficulty in deactivating them by pasteurization. Workers in the Cornell University Food Science Department studied the use of glass-immobilized trypsin for preventing oxidized flavor. Initial results were encouraging; immobilized trypsin can reduce the oxidative flavor of treated milk. Many "bugs" remain to be worked out. If too much trypsin is used, bitter and astringent flavors develop. Best operating conditions must be determined and the in-use stability of the supported enzyme evaluated. The immoblized enzyme is certainly expected to have advantages over free trypsin. Among these are the easy removal of the enzyme activity from the milk and the possibility of reuse which lowers the cost of the process.

Chapter Eight

Enzymes in Analysis

THE ROLE OF the analytical chemist is somewhat like that of a detective. He has basically two questions to answer: (1) what is the chemical composition of a given material, and (2) how much of the important ingredient (or ingredients) is there. The former question is answered by qualitative analysis, while the latter is determined by quantitative analysis. In the two chapters dealing with uses of enzymes in analysis, quantitative aspects are stressed. The section on enzymes in forensic analysis illustrates qualitative as well as quantitative functions of enzymes.

Analysis is essential in all endeavors involving chemical materials. The manufacturer analyzes raw materials purchased for use in his product to assure himself that his supplier has not (knowingly or unknowingly) misrepresented the composition of those materials. After preparing his product, he analyzes it to see that proper levels of important ingredients are present. He may store samples and analyze them periodically to determine their stability in storage. His product may be analyzed by state and federal regulatory agencies to protect the consumer from any erroneous claims on his part.

The forensic scientist is one who provides scientific evidence for use in law courts. He sometimes can confirm the guilt or innocence of an accused person by use of analytical techniques.

Medical diagnosis is frequently aided by analysis. The determination of high blood sugar may reveal a diabetic condition. Unusually high levels of enzymes in the blood or urine may help physicians to differentiate between various disease states and to estimate the severity of the disease.

Enzymic methods comprise a very small part of the armamentarium available to the analytical chemist. Nevertheless, enzymes have unique characteristics, and their progressively greater availability will doubtlessly increase their utilization in analytical procedures.

Probably the greatest advantages of enzymes in analysis are their sensitivity and their specificity. That is, they are able to detect very low levels of materials, and they often react with a particular molecule, while not affecting closely related ones. In addition, enzymic methods are accurate, relatively quick, and are not dependent on extreme conditions. They have been automated in recent years.

A variety of methods are available to the analyst using enzyme techniques. Most modern methods employ electronic instruments, which simplify the procedure and decrease the opportunity for human error. The usual method of enzymic analysis depends upon measuring the rate of decrease of a reactant, or the rate of increase of a product. If a reactant or a product (but not both) absorb light in a given region of the spectrum, that light absorbance can be measured and correlated with the concentration of reactants and products.

An example of this is provided by the conversion of uric acid to allantoin by uricase (uric acid oxidase), as shown in Equation 8–1.

Equation 8–1

$$\text{Uric acid and } O_2 \quad \xrightarrow{\text{Uricase}} \quad \text{Allantoin and } H_2O_2 \text{ and } CO_2$$

The reactant, uric acid, shows intense absorption of ultraviolet light at 290 nanometers. The product, allantoin, does not absorb light at that wavelength. By measuring the rate of decrease of absorption of light at 290 nanometers, the rate of uric acid conversion is measured and the uric acid concentration can be calculated.

Quite often, neither the reactant nor the product of a reaction absorbs light to a significant degree at a convenient wavelength. In this case, "coupled" enzymic reactions may be used to produce a product which does have significant absorption at a convenient wavelength. In many cases, the coupled reaction, which provides the light-absorbing product, uses a cofactor, such as nicotine adenine dinucleotide (NAD). When this cofactor is reduced, the product is called NADH. It has a strong absorbance in the ultraviolet at 340 nanometers. The following two equations (8–2 and 8–3) illustrate the point:

Equation 8–2

Glycerol and ATP

$$\text{Glycerol and ATP} \xrightarrow{\text{Glycerokinase; Mg}^{2+}} \text{L-glycerol-1-phosphate and ADP}$$

Equation 8–3

L-glycerol-1-phosphate and NAD

$$\text{L-glycerol-1-phosphate and NAD} \xrightarrow{\text{Glycerol phosphate dehydrogenase}} \text{Dihydroxyacetone phosphate and NADH}$$

None of the materials in Equation 8–2 absorbs light strongly in a convenient region of the spectrum. The product of the reaction, glycerol–1–phosphate, is therefore used as a substrate in a second reaction, which uses the cofactor, NAD. For each molecule of glycerol reacted, 1 molecule of glycerol–1–phosphate is formed in reaction 8–2. In reaction 8–3, one molecule of NADH is formed for every molecule of glycerol–1–phosphate reacted. The increase in absorption at 340 nanometers measures the increase in NADH. From this, the glycerol concentration is easily calculated.

A hundredfold increase in sensitivity in the measurement of a material can be obtained if a product of a reaction is fluorescent. Happily, NADH is such a compound. As we saw above, NADH absorbs light at 340 nanometers. When it fluoresces, it emits light at 450 nanometers. In practice, the reaction mixture is exposed to light at 340 nanometers. The detector, at 450 nanometers, is at right angles to the light source. The fluorescence at 450 nanometers can be used in calculations, just as absorbance at 340 nanometers is used. Whenever the cofactor NADH is involved in a reaction, fluorescent measurement is possible. Because of the greater sensitivity of fluorometric methods, smaller samples and smaller amounts of cofactors can be used in an analysis.

The analyses we will be concerned with in this chapter can be carried out using a spectrophotometer (an instrument which provides light of a known wavelength and measures the amount of absorbance at that wavelength) and a fluorometer (which provides light at one wavelength and measures emitted light at a second wavelength).

Other analyses may be followed manometrically (by measuring the volume of a gas liberated in a reaction), electrochemically (see

Chapter 9), thermochemically (by measuring the heat liberated or absorbed during a reaction), or radiochemically (using a radioactive substrate). Changes in pH resulting from a reaction can also be used to monitor the reaction.

Examples of Industrial Analysis Using Enzymes

The use of phosphates in detergents is undesirable because of the pollution problems they cause. Citrates are used in place of phosphates in detergents prepared for use in areas where phosphate-free detergents are mandatory. Unlike phosphates, citrates are rapidly biodegradable.

Since citrate detergents are likely to become progressively more prominent, an assay method for citrates in such preparations was needed. An enzymic method was devised by Taraborelli and Upton in Pfizer's quality control laboratory. A coupled sequence of reactions is used to give a fast, simple, and specific assay method. These reactions are shown in Equations 8–4 and 8–5.

Equation 8–4

$$\text{Citrate} \xrightarrow{\text{Citrate lyase}} \text{Oxaloacetate and Acetate}$$

Equation 8–5

$$\text{Oxaloacetate and NADH and H}^+ \xrightarrow{\text{Malate dehydrogenase}} \text{Malate and NAD}^+$$

NADH is destroyed in the "indicator reaction" of this coupled sequence. This destruction is measured by the decrease in absorption at 340 nanometers and the citrate concentration is easily calculated.

So far, we have only considered methods for measuring concentrations of substrates. It is also possible to measure concentrations of activators and inhibitors of enzymes.

An activator is a material, quite often a metal ion, which converts an "apoenzyme" (inactive) into an active enzyme. The activity will increase steadily until there is enough activator to convert all of the apoenzyme to the active form.

An example of this is provided by aminopeptidase derived from pig kidneys. Each molecule of this enzyme contains two zinc ions which are essential for enzymic activity. These ions can be removed, giving an inactive apoenzyme. Zinc ion thus is an activator of aminopeptidase, and the concentration of zinc in a solution can be measured by its ability to activate the apoenzyme.

In practice, the apoenzyme is added to a zinc-containing solution of known concentration. A proportionate amount of activity is restored. The activity is measured and the experiment is repeated with zinc solutions of varying concentrations. A graph of the increase in enzymic activity with increase in zinc concentration gives a straight-line plot when appropriate levels of zinc ion are used. Zinc solutions of unknown concentration can then be measured by carrying out the same experiment. The unknown zinc-ion concentration can be determined by comparison with the graph.

The insecticide DDT inhibits the enzyme carbonic anhydrase. This inhibition provides an enzymic method for measuring DDT. The reaction is shown in Equation 8–6.

Equation 8–6

$$H_2CO_3 \xrightarrow{\text{Carbonic anhydrase}} CO_2 \text{ and } H_2O$$

Carbonic acid → Carbon dioxide and Water

Carbonic anhydrase is assayed by measuring the carbon dioxide liberated from carbonic acid. DDT inhibits the enzyme and the rate of carbon dioxide evolution decreases proportionately. The experiment is carried out with various known concentrations of DDT. The percentage of inhibition caused by each concentration of insecticide is recorded and a plot of the percentage of inhibition versus the concentration of DDT is drawn. The experiment is repeated with DDT samples of unknown concentration, and the concentration is determined by comparing the percentage of inhibition with the graph.

FORENSIC ANALYSIS

The forensic scientist applies scientific methods to provide objective evidence in a court of law. His results are frequently crucial in establishing reasonable certainty of guilt or innocence.

Among the questions most frequently asked is "was this driver under the influence of alcohol?" A person is considered "under the influence" if the alcohol concentration in his blood exceeds a certain limit defined by state law.

One of the most elegant methods for answering this question

makes use of an enzyme—alcohol dehydrogenase. The reaction is shown in Equation 8–7.

Equation 8–7

$$CH_3CH_2OH \text{ and NAD} \xrightarrow{\text{Alcohol dehydrogenase}} CH_3CHO \text{ and NADH}$$

Ethyl alcohol   Acetaldehyde

This is another assay based on the absorbance (or fluorescence) of the reduced cofactor NADH. The rate of formation of NADH is measured at 340 nanometers and the alcohol concentration is calculated from this data.

The ideal source of the sample would be arterial blood reaching the brain. This is impossible, but alternative sources are considered quite reliable. These sources are the bloodstream, the urine, and expired breath. Four volumes of urine are considered to have the same alcohol content as three volumes of blood; and 2,100 volumes of air (breath) are considered to have the same alcohol content as one volume of blood.

It is not possible to analyze body fluids for alcohol directly, since all of these fluids contain other substances which would interfere with reactions of the alcohol. It is common practice to allow the alcohol in a small sample of blood or urine to evaporate into a buffered solution to be assayed using alcohol dehydrogenase and NAD. Air samples are usually assayed using nonenzymic methods.

One of the most crucial duties of the forensic scientist is the establishment of the personal identity of an accused culprit. If a sample of blood from the criminal is available for comparison with that of the accused, comparisons of blood proteins may be used to establish innocence or provide evidence of guilt. In using blood samples for forensic purposes, the blood type is established. If the two are identical in this respect, a study of isoenzymes in the blood is useful.

At least six enzymes are known to exist in isoenzymic forms in the blood of different individuals. These include cholinesterase, phosphoglucomutase, adenylate kinase, red cell acid phosphatase, glucose–6–phosphate dehydrogenase, and 6–phosphogluconate dehydrogenase. Isoenzymes differ in amino acid composition, although they catalyze the same reaction. Some have an excess of negatively charged amino acids giving the enzyme an overall negative charge; some have an excess of positively charged amino acids resulting in a positively charged enzyme;

some have equal numbers of positive and negative charges which cancel each other to give an "uncharged" molecule. Isoenzymic forms can be separated and identified by an electrical process known as electrophoresis.

In zone electrophoresis, the sample is applied in a narrow band to a support material such as a starch gel, cellulose, or polyacrylamide gel. It is applied approximately midway between buffer reservoirs in which electrodes are immersed. The support is in contact with the buffer in the reservoirs, so that a voltage applied across the elctrodes produces a current through the support. In response to the applied voltage, those proteins having an overall positive charge at the pH employed will be attracted toward the negative electrode; negatively charged proteins will be attracted toward the positive electrode. Proteins, which are uncharged at the pH used, do not move. If two or more isoenzymes have differing degrees of excess charge, they will be attracted toward the electrode of opposite charge at rates which reflect the differing degrees of charge. For instance, an enzyme having 10 excess negative charges would move toward the positive electrode faster than one having 5 excess negative charges, but slower than one having 15 excess negative charges. In a given period of time, the enzyme with 15 excess charges would move most; the one with 5 would move least.

If this technique is applied to the blood sample left by the culprit and to a sample taken from the accused, and it is found that all pertinent serum isoenzymes are identical in both samples, there is good evidence of guilt. A finding of gross differences is good news for the defense.

Enzymic analysis is also useful in cases of alleged rape. Human seminal fluid is uniquely rich in the enzyme acid phosphatase. To extract the enzyme, damp filter paper is pressed against the article suspected of containing semen. The paper is then treated with an acidic solution containing a substrate for the enzyme reaction—sodium alpha-naphthyl phosphate—and a "coupler." If the article contained acid phosphatase, the enzyme would liberate alpha-naphthol from the substrate. Alpha-naphthol reacts with the "coupler" to give a highly colored product. Electrophoresis can be used to determine if the acid phosphatase is the isoenzyme formed in the prostate and found in semen, or if it is derived from some other source. A positive finding in these tests strongly suggests the presence of semen. Such a finding is usually followed by looking for spermatozoa (sperm cells) in the stain under a microscope. Finding spermatozoa is absolutely conclusive proof of some sort of sexual relations, since they occur only in seminal fluid.

Another possible use of electrophoresis in rape cases was revealed in a report attributed to Drs. G. F. Sensabaugh and E. F. Blake of the University of California at Berkeley. The report describes "genetic fingerprints" which are revealed by electrophoresis. Three different diaphorase patterns were found in sperm samples, but nowhere else in the body. Along with other substances that are found to varying extents in the sperm and other tissues of different individuals, the diaphorase isoenzymes make it possible to distinguish semen samples from two individuals with a probability of about 95%.

Regulatory Uses of Enzymes

The regulatory chemist serves as a sort of quality control scientist protecting the interests of the consumer. Just as a quality control chemist in a commercial operation uses analysis to protect his company, the regulatory chemist uses analysis to protect his sponsors—the taxpaying consumer. Techniques used by the regulatory scientist include all known analytical methods, but the most important are those recommended by the Association of Official Analytical Chemists (AOAC). Although the use of enzymes has not been fully exploited by this group, there are some noteworthy uses of enzymes in regulatory analysis.

In a highly agricultural state, such as Nebraska, the State Agriculture Laboratory carries out analyses designed to protect the rancher and, ultimately, the consumer. The analysis of animal feeds illustrates a procedure which employs an enzyme, although not directly as an analytical reagent. Animal feeds contain a certain level of protein, which is specified in the guarantee by the manufacturer. Ruminant animals, such as cattle, can use a fairly large proportion of urea, which is much cheaper than protein. The urea is used by these animals to build body protein. (Urea has the structure $H_2N–CO–NH_2$.) In high concentrations, urea is toxic, even to ruminants. The State Agriculture Laboratory determines the "total protein" in a sample by decomposing the sample with acid. In this procedure, all nitrogen-containing compounds are broken down to yield ammonia. A second sample is decomposed by the enzyme urease. This enzyme does not decompose any protein, but breaks down all urea to carbon dioxide and ammonia. The ammonia is collected and measured. Ammonia formed in the acid-catalyzed reaction measures the "total protein"—that is, true protein plus urea. That formed in the enzymic reaction measures only the urea—the so-called nonprotein nitrogen (NPN). This is subtracted from the "total protein" to get "true protein." If total protein is too low, or if NPN is too high, the sample can be declared in violation of the guarantee and its sale in the state stopped.

Just as acid phosphatase is used in the dairy industry to test the efficiency of pasteurization, so is it also used by regulatory agencies. Yet the test is not infallible. It is based on the fact that acid phosphatase is more resistant to heat than are pathogenic bacteria. Thus, if the acid phosphatase activity in milk is reduced below a critical level, the milk is assumed to be adequately pasteurized. But a mixture of overheated and underheated milk could have the same level of acid phosphatase activity and still contain pathogenic bacteria, which would continue to multiply. Thus, a sample of milk which seems safe may, in fact, be quite hazardous.

Conversely, it is known that acid phosphatase, which has been deactivated by pasteurization, is often reactivated after the process is completed. In this case, a high level of acid phosphatase may be found despite complete removal of pathogens.

A test has been devised to differentiate reactivated acid phosphatase from that due to raw or underpasteurized milk products. A product capable of reactivation, when stored with magnesium chloride at 34° C, will show a tenfold increase in acid phosphatase activity when compared with the same product stored without the magnesium salt. The phosphatase activity in underpasteurized products does not show such an increase in activity when stored with magnesium chloride.

Proper refrigeration of samples is critical if this test is to be applied to distinguish regeneration from inadequate pasteurization. If a sterile product is allowed to warm up to around room temperature and left at that temperature for as long as 2 hours, the test cannot be considered reliable. Microbiological tests for pathogens would then be needed.

ENZYMES IN CLINICAL ANALYSIS

Clinical chemistry is a field in which enzymic analysis has achieved considerable importance. The clinical chemist, on the request of a physician, analyzes body fluids or tissues to determine the concentration of materials which will aid the physician in his diagnosis and prognosis. Quite often, it is a determination of an enzyme itself which is used to diagnose a pathological condition or to confirm a diagnosis. In the following pages, we shall see how enzymic analysis is useful in diagnosing disease states, including such life-threatening conditions as cancer, heart defects, and diabetes.

In a healthy individual, most enzymes are synthesized inside a cell, utilized inside that cell, and ultimately decomposed inside that same cell. There are, of course, exceptions such as the digestive enzymes formed in the pancreas for use in the small intestine. But, by and large, the enzymes stay, not only in the cell where they were made, but even in the subcellular particles of their origin. There is a very small level of these cell-bound enzymes in the blood of healthy individuals. These are believed to be in the blood as a result of normal cell breakage and cell membrane defects. They provide the so-called "normal reference level." Blood levels of enzymes greatly in excess of the normal reference level are generally associated with pathological conditions and are therefore useful in diagnosing such conditions.

The enzymes found in the serum in diseased states are conveniently considered in two groups—the enzymes of metabolism, which are found in all tissues, and the so-called "organ-specific enzymes," which are found in relatively few organs.

Enzymes in the Diagnosis
of Cardiovascular Trauma

It is fitting to consider cardiovascular problems first, not only because they are the leading killers in our society, but also because some interesting techniques are involved in their diagnosis. Although eight different enzyme systems have been found to have some merit in diagnosing cardiovascular diseases, three of these are of particular value and interest. These three will be considered: creatine phosphokinase (CPK), lactic dehydrogenase (LDH), and "hydroxybutyrate dehydrogenase" (HBD).

Creatine phosphokinase occurs in significant concentrations in the brain and in the muscle, including heart muscle. It is an organ-specific enzyme for heart disease if muscle diseases (as muscular dystrophy), brain damage, and hypothyroidism are ruled out. Following myocardial infarction (a blockage in a blood vessel of the heart), the serum level of creatine phosphokinase increases within 3 hours and reaches a peak after 36 hours. Peak values range from 5 to 25 times the normal reference level. Generally, CPK returns to normal levels after 4 days. If CPK remains high after this period, the prognosis is poor. Drugs, such as opiates, penicillin, anticoagulants, or alcohol injected into a muscle can raise serum CPK levels.

A three-step "coupled sequence" of enzymic reactions is used to measure the activity of CPK in serum. These reactions are illustrated in Equations 8–8, 8–9, and 8–10.

Equation 8–8

$$\text{Creatine phosphate and ADP} \xrightarrow{\text{CPK}} \text{Creatine and ATP}$$

Equation 8–9

$$\text{Glucose and ATP} \xrightarrow{\text{Hexokinase}} \text{Glucose-6-phosphate and ADP}$$

Equation 8–10

Glucose-6-phosphate and NADP
$$\xrightarrow{\text{G-6-PDH*}} \text{6-phosphogluconate and NADPH}$$

---

*G-6-PDH is the enzyme glucose-6-phosphate dehydrogenase.

The cofactor, NADPH (nicotine adenine dinucleotide phosphate), like NADH, absorbs ultraviolet light at 340 nanometers. Thus, we have a coupled series of three enzymic reactions for determining the activity of CPK by measuring the rate of formation of NADPH.

Isoenzymes in Diagnosis

The diagnostic use of enzymes suffers from a lack of complete organ specificity. Creatine phosphokinase activity has been seen to be elevated in cases of myocardial infarction. There are reports of CPK elevation in other diseases not involving the heart. These include muscular dystrophy, tetanus, and strokes. How, then, do we know whether an elevated CPK is due to heart, muscle, or brain disease? A separation of isoenzymes helps to answer this.

There are three isoenzymic forms of CPK. Heart muscle contains CPK isoenzymes designated CPK3 and CPK2. Skeletal muscle contains CPK3 and minimal amounts of CPK2, while the nervous system and brain contain CPK1. Thus, by determining which CPK isoenzyme is elevated in the blood, an additional clue to the source of the enzyme is obtained. The most sensitive and specific laboratory procedure for identifying necrosis (death due to a lack of blood supply) of heart muscle is the presence of CPK2 in the patient's serum. This isoenzyme determination supplements the electrocardiogram in the assessment of possible necrosis of heart muscle.

Lactic dehydrogenase (LDH), the first isoenzyme system used in diagnosing myocardial infarction, exists in five forms. Normal serum shows a predominance of LDH2; LDH1 and LDH3 are of about equal concentration, somewhat lower than LDH2; LDH4 and LDH5 are about equal and considerably lower than the others. An increase in LDH1 in the serum, making its activity greater than that of LDH2, is seen in acute myocardial infarction, muscular dystrophy, and a particular form of kidney disease. The LDH1 level increases about 12 to 18 hours after the infarction and persists for about a week.

Activity of LDH can be determined by measuring the NADH formed in the reaction shown in Equation 8–11.

Equation 8–11

$CH_3CHOHCOOH$ and NAD
Lactic acid

$$\xrightarrow{\text{Lactate dehydrogenase}} CH_3COCOOH \text{ and } NADH$$
Pyruvic acid

"Hydroxybutyrate dehydrogenase" (HBD) is, in fact, a mixture of LDH1 and LDH2. It was discovered that these two isoenzymes of LDH, but not the other three, catalyze the reaction shown in Equation 8–12.

Equation 8–12

$$CH_3CH_2COCOOH \text{ and } NADH \xrightarrow{\text{HBD}} CH_3CH_2CHOHCOOH \text{ and } NAD$$
Alpha-ketobutyric acid   Alpha-hydroxybutyric acid

By using a different substrate, alpha-ketobutyric acid instead of lactic acid, it is thus possible to measure the activity of two isoenzymes without interference from the other three.

Lactate dehydrogenase isoenzymes 4 and 5 are helpful in diagnosing liver disorders. Still another method for distinguishing LDH isoenzymes of the heart from those of the liver make use of heat inactivation of the enzymes. If the activity of LDH in the serum is measured, the serum incubated at 65°C for 30 minutes, and the LDH activity again measured, the activity will have decreased due to protein denaturation. But the enzyme from the heart is more stable to heat than is that from the liver. If the enzyme in the serum came from heart muscle, about half of the original activity will remain; if it came from the liver, only about 10% will remain.

ENZYMES IN DIAGNOSIS OF CANCER

Acid phosphatase is one of the most useful enzymes for diagnosing neoplastic diseases (cancers). This enzyme is found in very high concentrations in the prostate gland and is considered organ specific. Yet, Bodansky found that serum acid phosphatase was elevated in the vast majority of patients having cancer of the prostate which metastasized (spread) to the bone, but only in relatively few patients whose prostate cancer had not spread. Thus, elevated serum acid phosphatase is more valuable in confirming the spread of cancer than in diagnosing cancer of the prostate.

Inhibitors have been found to be of value in making serum acid phosphatase useful in diagnosing prostate cancer before the cancer spreads and before the serum enzyme activity rises. The salt L–tartrate inhibits acid phosphatase derived from the prostate, but not that found in other tissues. Green and co-workers at the New England Medical Center found elevated prostatic acid phosphatase in the serum of 44 of 45 patients with cancer of the prostate at a time when the total acid phosphatase was elevated in only 24 of these cases. Since cancer which has spread causes far more deaths than localized cancers, this is an important observation. It permits diagnosis before spreading has occurred.

Alkaline phosphatase is often elevated in the serum of patients having tumors involving bone or liver. Three other enzymes—leucine aminopeptidase, 5′–nucleosidase, and gamma-glutamyl transpeptidase—are also elevated in the case of liver disease, but none of these can be affected by bone diseases. They therefore serve to indicate whether elevated alkaline phosphatase is from bone or liver.

The activities of many widely occurring enzymes are also elevated in cancerous states, but these are not valuable in diagnosis. They are useful in following the course of the disease and the growth or regression of the tumor. Thus, they assist in determining whether the therapy is effective or not. Some attempts have been made to increase their value as diagnostic aids by measuring isoenzymes. Aldolase has shown some promise in this regard.

ENZYMIC DIAGNOSIS OF DIABETES

In the diabetic person, glucose accumulates in the blood due to accelerated glucose formation coupled with retarded glucose utilization;

the ability of the kidneys to excrete glucose is impaired. Arterial disease frequently accompanies diabetes. Severe acidosis develops in many diabetic persons due to hepatic (liver) formation of acetoacetic acid, beta-hydroxybutyric acid, and, sometimes, lactic acid.

There are various forms of diabetes—diseases of widely varying severity. The juvenile form is much more serious than the adult form which attacks late in life. There is a condition known as "chemical diabetes" in which the victim has abnormal blood glucose in a glucose tolerance test, but has no symptoms, and has a normal level of glucose after fasting. The symptomatic diabetic complains of increased urine, increased appetite and thirst, and has elevated levels of glucose in the blood and urine.

The test most generally used for measuring the blood sugar (glucose) level for diabetes diagnosis and control is an enzymic process using glucose oxidase, peroxidase, and a material which can form a dye. The sample (serum, plasma, or whole blood) is deproteinized by a mixture of barium hydroxide and zinc sulfate. It is then incubated with glucose oxidase, peroxidase, and ortho-dianisidine. The process is illustrated in Equations 8–13 and 8–14. Glucose is oxidized

$$\textbf{Equation 8–13}$$

$$\text{Glucose and } H_2O \text{ and } O_2 \quad \xrightarrow{\text{Glucose oxidase}} \quad \text{Gluconic acid and } H_2O_2$$

$$\textbf{Equation 8–14}$$

$$H_2O_2 \text{ and Ortho-dianisidine} \quad \xrightarrow{\text{Peroxidase}} \quad \text{Colored compound and } H_2O$$

and hydrogen peroxide is formed. Peroxidase decomposes the peroxide-liberating oxygen which reacts with ortho-dianisidine giving an amber-colored product. This is treated with sulfuric acid to give a rose pink color which is measured in a spectrophotometer at 540 nanometers.

Another enzymic system has found some use in measuring blood glucose. The sample is incubated with a mixture containing hexokinase and a cofactor, adenosine triphosphate (ATP), and glucose–6–phosphatase dehydrogenase and its cofactor, nicotine adenine dinucleotide phosphate (NADP). The process is illustrated in Equations 8–15 and 8–16. The NADPH is measured at 340 nanometers (ultraviolet) and glucose is easily calculated. This method has some advantage in that deproteination is not necessary.

Equation 8 15

Glucose and ATP $\xrightarrow{\text{Hexokinase}}$ Glucose-6-phosphate and ADP

Equation 8–16

Glucose-6-phosphate and NADP $\xrightarrow{\text{G-6-PDH}^*}$ 6-phosphogluconate and NADPH

---

*G-6-PDH is the enzyme, glucose-6-phosphate dehydrogenase.

There are many other enzymic analyses which are important in clinical chemistry. For instance, they are useful is diagnosing kidney problems and liver diseases, such as hepatitis and obstructive jaundice. Space limitations make it impossible to consider these further.

Additional analytical uses of enzymes are described in the next chapter and in Chapter 15 in the section on "enzyme inhibitors in war and agriculture."

Chapter Nine
Immobilized Enzymes in Analysis

IN THIS CHAPTER we continue our discussion of analytical uses of enzymes, but the topic is now limited to immobilized enzyme applications. We have seen how an enzyme activator, such as a metal ion, can be assayed by use of the enzyme which it activates. This same technique can be used with immobilized enzymes.

Stone and Townshend in England have described an assay of "ultratrace" levels of copper ion, by use of immobilized polyphenol oxidase. The enzyme was chemically bonded to the polyacrylamide derivatives known as "Enzacryls."

Copper was removed from the enzyme by treating it with a solution containing cyanide ion. The cyanide competes with the enzyme for the copper, and conditions are rigged to ensure the victory of the cyanide in the competition. The copper was removed from the soluble enzyme in a matter of days; it was removed from the immobilized enzyme in minutes.

The resulting "apoenzyme" can be reproducibly and selectively reactivated by incubation with solution containing from 25 to 200 nanograms of copper ion (a nanogram is one-billionth of a gram). The extent of activation of the enzyme measures the concentration of copper ion. Both soluble and insolubilized polyphenol oxidase take up the copper and regain activity; the insoluble material takes it up at a markedly lower rate. There is a linear relationship between the copper-ion concentration and the degree of activation of the enzyme. Thus, a calibration curve based on a small number of solutions of known copper-ion concentration can be drawn and used to determine unknown copper-ion concentrations.

The insolubilized enzyme can be used several times without excessive loss of enzyme activity. The method provides a highly selective means of detecting and measuring nanogram levels of copper ion with "reasonable accuracy and precision."

Reagentless Assays

The work load in clinical laboratories and other analytical laboratories has increased markedly in recent years. A good deal of the drudgery in this work involves the preparation of reagents—the chemical solutions required for the various procedures. These materials must be carefully weighed, dissolved, and standardized before they can be used to measure materials of interest. In order to eliminate some of this drudgery, extensive research has been carried out in an effort to devise "reagentless assays." Immobilized enzymes are useful tools in this research.

Measurement of Uric Acid
by Use of Urate Oxidase

The measurement of the concentration of uric acid in biological fluids, such as serum and urine, is of fundamental importance in the diagnosis and study of diseases involving malfunctions of purine metabolism. Gout is a classic example of such a disease. Enzymic methods have achieved prominence because of their specificity and sensitivity. In addition, they do not require removal of protein, as alternative methods do. Dritschilo and Weibel, at the University of Pennsylvania, have described an essentially reagentless assay of uric acid using immobilized urate oxidase.

Until now, we have been concerned with assays based on measuring the rates of enzyme-catalyzed reactions. This uric acid determination uses the "end-point" method, that is, the enzyme affects a total conversion of the substrate and this total conversion, not its rate, is measured. Advantages of this approach are said to include high reproducibility, strict linearity, and simplicity.

The enzyme was chemically bonded to porous glass and 0.8 milliliters of the adduct was placed in a small column. Fluids were saturated with oxygen and passed through the column at flow rates of 3 to 6 milliliters per minute. The enzyme catalyzes the reaction shown in Equation 9–1.

Equation 9–1

$$\text{Uric acid and } O_2 \text{ and } H_2O \quad \xrightarrow{\text{Urate oxidase}} \quad \text{Allantoin and } H_2O_2 \text{ and } CO_2$$

After this reaction in the column, the fluid goes to the detector, which is an oxygen-sensitive electrode. This device measures the depletion of oxygen caused by the reaction. As the equation shows, 1 molecule of oxygen is used for each molecule of uric acid present. Thus, by measuring the total decrease of oxygen, the total level of uric acid is determined.

Fresh uric acid standards were prepared daily to check the calibration of the device and the activity of the immobolized enzyme. Data were collected over a period of 3 months using the same enzyme column. It was used 2 to 4 days per week and was left in place at room temperature while not in use. A typical assay time was 2 minutes.

A number of urine samples were assayed for uric acid using the soluble enzyme and the glass-enzyme adduct. There was good agreement between the two methods, with an average standard deviation of 3% to 4% by either method.

The immobilized enzyme was usable for 90 days at room temperature and could be stored for 6 months at 4°C without appreciable loss of activity. Thus, the cost of the enzyme per assay is reduced to the point of insignificance. The determinaton is very fast, and, except for a buffer solution, which is mixed with the biological fluid, it is reagentless.

Many other compounds can be assayed using the same device, merely by replacing the glass-supported urate oxidase with another immobilized oxidase, which is specific for the substrate one wishes to measure. For instance, a column of glass-supported galactose oxidase might be used to measure galactose.

This represents one approach to the development of a reagentless assay. Another approach is based on the combination of two rapidly developing areas of technology—immobilized enzymes and ion-selective electrodes. The product of this combination is known as the "enzyme electrode."

Several years before the advent of the first enzyme electrode, Dr. L. C. Clark of the University of Cincinnati combined an electrochemical oxygen sensor with oxidases in solution and demonstrated the advantages of the combination. It seems likely that Dr. Clark's creativity hastened the invention of enzyme electrodes.

The first enzyme electrode was described by Drs. S. J. Updike and G. P. Hicks at the University of Wisconsin. An oxygen-sensitive electrode was made sensitive to changes in glucose concentration by covering the electrode with a gelatinous membrane consisting of glucose oxidase entrapped in polyacrylamide gel. The immobilized enzyme was held on the electrode by a thin nylon net impregnated with silicone plastic cement. The oxygen electrode measures the flow of oxygen through a membrane. The current put out by the electrode is proportional to the oxygen pressure. In a solution containing glucose, the oxygen pressure is reduced in the enzyme gel by the reaction shown in Equation 9–2. Under appropriate conditions, there is a linear relationship between the glucose concentration in solution and the decrease in

Equation 9–2

$$H_2O + \text{Glucose and } O_2 \xrightarrow{\text{Glucose oxidase}} \text{Gluconic acid and } H_2O_2$$

oxygen pressure. The device is calibrated by use of a series of glucose solutions of known concentration. Each of these is measured using the enzyme electrode, and a graph is drawn showing the current reading as a function of glucose concentration. By use of this graph, the glucose concentration of unknown solutions can be determined from the current reading obtained from the enzyme electrode in those solutions.

Dr. George G. Guilbault of the University of New Orleans has been responsible for a good deal of the development of the enzyme electrode. He has described three variations of the glucose oxidase electrode based on different modes of immobilization of the enzyme. In his type 1 electrode, a thin layer of chemically bonded glucose oxidase is trapped between the electrode and a cellophane film. Type 2 is similar to the Updike and Hicks electrode using polyacrylamide-entrapped enzyme. Type 3 uses glucose oxidase powder between the electrode and a cellophane film. (Cellophane is a "semipermeable membrane; it permits free passage of small ions and molecules, but holds the large protein molecules in place.) The chemically bonded enzyme electrode was the most stable and the glucose oxidase powder was least stable. The electrodes were found to have good accuracy, precision, sensitivity, and selectivity in measuring glucose in very small (0.1 milliliter) blood samples. The method is said to be simple, rapid, and inexpensive enough for use in physicians' offices and small hospitals or clinical laboratories where rapid measurements are needed. The only reagent needed is a phosphate buffer.

Nilsson and co-workers in Sweden used a pH-sensitive electrode to prepare enzyme electrodes based on glucose oxidase, penicillinase, and urease. As shown in Equations 9–3, 9–4, and 9–5, the first two reactions produce acid and the third consumes it. The electrode senses the changes in acidity caused by the enzyme-catalyzed reactions.

Equation 9–3

$$\text{Glucose and } O_2 \text{ and } H_2O \xrightarrow{\text{Glucose oxidase}} \text{Gluconic acid and } H_2O_2$$

Equation 9–4

$$\text{Penicillin and } H_2O \xrightarrow{\text{Penicillinase}} \text{Penicilloic acid}$$

Equation 9–5

$$\text{Urea and } H^+ \text{ and } H_2O \xrightarrow{\text{Urease}} 2NH_4^+ \text{ and } HCO_3^-$$

The method based on the pH-sensitive electrode is versatile because many enzyme-catalyzed reactions use or produce hydrogen ions. Thus, it should be possible to prepare enzyme electrodes to measure concentrations of substrates which are converted in any enzymic reaction producing or using hydrogen ions (acid).

The ammonium ion is another chemical species produced in many enzyme-catalyzed reactions. Guilbault and others have prepared enzyme electrodes using a number of these ammonium ion-producing enzymes. The first of these was the urease electrode.

Nylon netting was soaked in a solution containing acrylamide, urease, and a cross-linking agent. The net was placed over a cation-sensitive electrode and the acrylamide was polymerized. (The cation-sensitive electrode responds to sodium, potassium, and hydrogen ions as well as the ammonium ion.) The polyacrylamide film was protected by dialysis paper—another semipermeable membrane. In order to use this electrode for measuring urea in blood or urine, it was necessary to combine the urease electrode with a glass-reference electrode and to remove interfering ions by use of ion-exchange resins.

We have seen the importance of L–amino acids in the food industry. These are the building blocks of proteins. Guilbault described an enzyme electrode for use in measuring the concentrations of these amino acids. This enzyme electrode uses two enzymes, L–amino acid oxidase and catalase, as shown in Equations 9–6 and 9–7.

Equation 9–6

$$\text{RCH NH}_3^+ \text{ COO}^- \text{ and H}_2\text{O and O}_2 \xrightarrow{\text{Amino acid oxidase}} \text{RCOCOO}^- \text{ and NH}_4^+ \text{ and H}_2\text{O}_2$$

Equation 9–7

$$2\,\text{H}_2\text{O}_2 \xrightarrow{\text{Catalase}} 2\,\text{H}_2\text{O and O}_2$$

As Equation 9–6 shows, ammonium ions are produced by the amino acid oxidase reaction. The added catalase decomposes the hydrogen peroxide formed in that reaction and provides more oxygen to keep the reaction going. It should also be recalled that enzyme-catalyzed reactions are reversible; they do not always go to completion, but frequently go to some equilibrium state, in which reactants and products remain in a constant ratio. If one of the products of a reaction can be removed, the equilibrium composition will shift to form additional product to replace that which was removed. (This is known as LeChatlier's principle.) Catalase, by removing hydrogen peroxide, not only supplies more oxygen, but also favors more product formation because it removes one of the products of the first reaction. Guilbault found an improvement in the response and the stability of the electrode after the inclusion of catalase.

Guilbault also prepared enzyme electrodes using D–amino acid oxidase for measuring concentrations of the "unnatural" D–amino acids. The stability of the electrodes was disappointing until it was discovered that the enzyme was losing a cofactor when stored in buffer overnight. This cofactor, flavin adenine dinucleotide (FAD), is held to the enzyme very loosely and was leached out by the buffer. When the electrodes were stored in a solution of FAD overnight, they were stable for at least 21 days.

The amino acids asparagine and glutamine are of special interest because they seem to be essential nutrients for some cancer cells, although they are not essential for healthy cells. Electrodes for measuring these amino acids have also been prepared by Guilbault's group. Both of these devices use the Beckman cation-sensitive electrode to detect ammonium ions, which are formed in the enzymic conversions illustrated in Equations 9–8 and 9–9.

To prepare the glutamine electrode, a piece of nylon stocking soaked with glutaminase solution was placed over the electrode and a film of dialysis paper was used to trap the enzyme. This "liquid mem-

Equation 9-8

$$H_2N\overset{\overset{\displaystyle O}{\|}}{C}CH_2\overset{\overset{\displaystyle NH_2}{|}}{C}HCOOH \text{ and } H_2O \text{ and } H^+$$
Asparagine

$$\xrightarrow{\text{Asparaginase}}$$

$$HOOCCH_2\overset{\overset{\displaystyle NH_2}{|}}{C}HCOOH \text{ and } NH_4^+$$
Aspartic acid and Ammonium ion

Equation 9-9

$$H_2N\overset{\overset{\displaystyle O}{\|}}{C}CH_2CH_2\overset{\overset{\displaystyle NH_2}{|}}{C}HCOOH \text{ and } H_2O \text{ and } H^+$$
Glutamine

$$\xrightarrow{\text{Glutaminase}}$$

$$HOOCCH_2CH_2\overset{\overset{\displaystyle NH_2}{|}}{C}HCOOH \text{ and } NH_4^+$$
Glutamic acid and Ammonium ion

brane" type of electrode is useful for about 12 hours, but loses sensitivity rapidly after that. An attempt to prepare a more stable electrode by use of polyacrylamide-entrapped glutaminase produced an electrode completely insensitive to changes in glutamine concentration. Apparently, the enzyme was deactivated in the polymerization reaction.

The liquid membrane asparaginase electrode gave a good response, but it also was unstable, losing about half of its response in 4 days. Polyacrylamide-entrapped asparaginase gave an electrode that was quite stable for more than 3 weeks.

A constant problem, when using the Beckmann cation-sensitive electrode to measure ammonium-ion formation, is the interference due to potassium, sodium, and hydrogen ions. As we mentioned earlier, the technology of ion-selective electrodes is developing rapidly at present. New ion-selective electrodes are being developed with increased sensitivity and selectivity; therefore, interferences are being reduced. Newer enzyme electrodes have been based on electrodes which are specific for gases such as carbon dioxide and for ions such as ammonium ion, iodide ion, and cyanide ion. Dr. Gary A. Rechnitz of the State University of New York at Buffalo deserves a good deal of credit for the development of modern ion-selective electrodes.

A new method for immobilizing enzymes on the surface of ion-selective electrodes has also improved the enzyme electrode. A French group described the method in which a mixture of an enzyme and bovine serum albumin (which acts as a "protective" protein) are mixed together and cross-linked on the surface of the electrode. As in most enzyme crosslinking procedures, glutaraldehyde was used for bonding.

More recently, Calvot, Thomas, and co-workers at the Enzyme Technology Laboratory at Compiegne, France, described an enzyme electrode, which uses a magnetic enzyme-containing membrane. Decarboxylases (enzymes which release carbon dioxide from carboxylic acids) specific for the amino acids tyrosine, phenylalanine, and lysine were used to make enzyme electrodes specific for each of these amino acids. A decarboxylase was cross-linked with an inert protein such as albumin, and magnetic ferrite particles were entrapped within the membrane structure. Glutaraldehyde was used to cross-link the proteins. Mechanical properties of the membrane were similar to those of cellophane and the enzymic activity was stable for weeks. The presence of the magnetic particles makes it possible to attach the enzyme-containing membrane to a CO_2-sensitive electrode bearing a cylinder magnet. The decarboxylase membrane is situated on the outer surface of the gas-permeable membrane.

As shown in Equations 9–10, 9–11, and 9–12, each of the enzymes in this study releases carbon dioxide from a specific amino acid.

The carbon dioxide electrode senses the carbon dioxide released in the reaction. The decarboxylases are highly specific for their particular substrate and the gas-sensitive CO_2 electrode is much more selective than the ion-specific electrodes. Therefore, interferences are minimized. Calibration curves are plotted by measuring samples of known concentrations of the amino acid. Concentrations of unknown solutions are determined by use of the calibration curves. The electrode provides a continuous response as long as it is in contact with a substrate-containing solution. Thus, it can be used in individual assays or in flowing streams to monitor the substrate concentration.

Cordonnier, Thomas, and others in the same French laboratory used the same magnetic membrane technique to prepare the first enzyme electrode using two different enzymes which react in a coupled sequence, with the second reaction providing a detectable species. Each of the electrodes used glucose oxidase. In different electrodes, this was coupled with lactase, maltase, and saccharase—enzymes which form glucose in their reactions, as seen in Equations 9–13, 9–14, and 9–15. Glucose, in turn, is converted to gluconic acid and hydrogen peroxide by oxygen in the presence of glucose oxidase. This is shown in Equation 9–16.

The magnetic, two-enzyme membrane was attached to an elec-

Equation 9–10

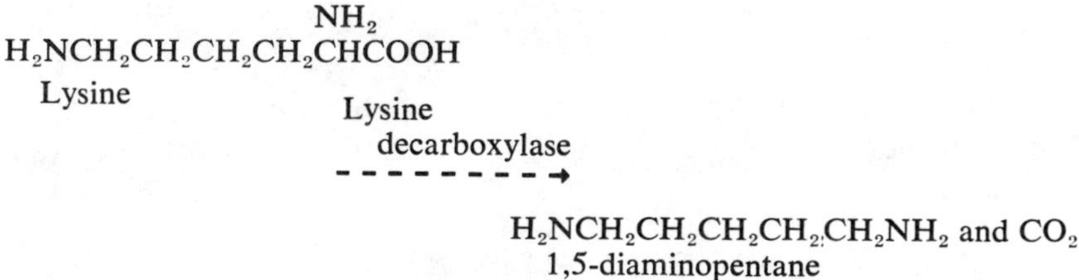

Tyrosine → Tyrosine decarboxylase → Tyramine and CO_2

Equation 9–11

$$H_2NCH_2CH_2CH_2CH_2CHCOOH$$
Lysine

Lysine decarboxylase
$$\longrightarrow$$

$$H_2NCH_2CH_2CH_2CH_2CH_2NH_2 \text{ and } CO_2$$
1,5-diaminopentane

Equation 9–12

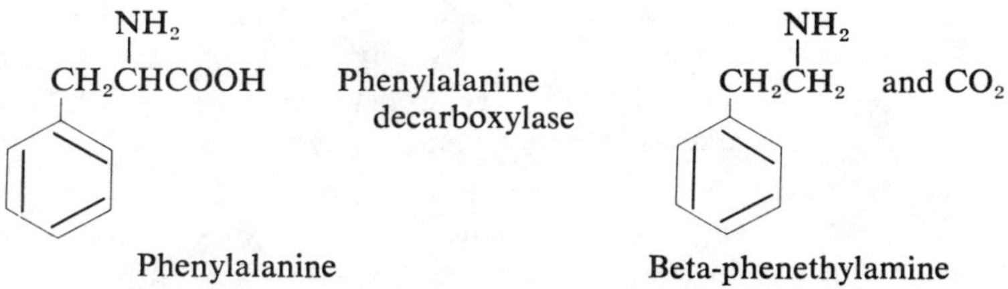

Phenylalanine → Phenylalanine decarboxylase → Beta-phenethylamine and CO_2

trode, which was sensitive to changes in oxygen levels. For each glucose molecule 1 molecule of oxygen is used and each glucose molecule is equivalent to 1 molecule of lactose or sucrose or to ½ maltose molecule. Calibration curves were plotted using substrate solutions of known concentration. A graph of the oxygen level as a function of substrate con-

Equation 9–13

$$\text{Lactose} \xrightarrow{\text{Lactase}} \text{Glucose and Galactose}$$

Equation 9–14

$$\text{Maltose} \xrightarrow{\text{Maltase}} \text{2-glucose}$$

Equation 9–15

$$\text{Sucrose} \xrightarrow{\text{Saccharase}} \text{Glucose and Fructose}$$

Equation 9–16

$$\text{Glucose and } O_2 \xrightarrow{\text{Glucose oxidase}} \text{Gluconic acid and } H_2O_2$$

centration gives a straight line over a reasonable concentration range. The electrodes provide a useful tool for measuring levels of important substrates.

FLUOROMETRIC ANALYSIS ON SOLID PADS

We have already mentioned the extreme sensitivity of fluorometric techniques of analysis. Guilbault has used this sensitive method in another approach to the development of reagentless assays.

In this approach, all of the reagents are dissolved in water or a suitable buffer, and the solution is deposited on a cured silicone rubber pad. The solvent is evaporated off, usually by lyophilization (freeze-drying), leaving a deposit of the reagent chemicals on the pad. The reagent pads may be prepared in large quantities and stored in the dry state, usually in a refrigerator.

To use the pads, the technician simply removes a pad from its container, allows it to warm to room temperature (about 30 minutes), adds water to dissolve the reagents, and finally adds his sample to the pad.

The pad is placed in the fluorometer and the rate of appearance, or disappearance, of fluorescence is charted on an electronic recorder. The result is determined by comparing this rate with a calibration curve. A typical assay can be completed in less than 5 minutes.

In addition to this impressive rapidity, there are several other significant advantages to this method. The reaction is carried out on a "micro" scale; that is, very small quantities of materials are used. For instance, a 3 microliter (approximately 0.000003 quart) sample of serum is adequate for an assay. Similarly, small quantities of reagents are used in preparing the pads. Quite frequently, this results in a significant cost reduction. The fluorometric method is extremely sensitive and is useful over an extended range of concentrations.

The method is relatively insensitive to temperature. This results from the fact that the silicone rubber pad does not conduct heat. As long as the sample is used at the same temperature as was used to prepare the calibration curve (usually room temperature, 25° C), the environmental temperature does not matter.

The reagents used in many enzyme assays are quite unstable and must be prepared fresh daily. But, when they are stored in the dry state on these pads, they can be kept for weeks or months with no deterioration.

Substrates of clinical importance have been of the greatest interest in developing the method. The first material to be assayed by this technique was the enzyme cholinesterase. The nonfluorescent substrate, N–methylindoxyl acetate, is converted to fluorescent N–methylindoxyl alcohol by the enzyme. Cholinesterase can be detected at low levels by this technique. Calibration curves are linear in the concentration range from 10^{-6} to 10^{-2} (0.000001 to 0.01) units per milliliter with excellent accuracy and precision. The enzyme cholinesterase is further discussed in Chapter 15.

Recently (1975), Guilbault's group applied the method to serum glutamate oxaloacetate transaminase (SGOT), serum glutamate pyruvate transaminase (SGPT), and hydroxybutyrate dehydrogenase (HBD). The first two of these enzymes are important in the diagnosis of liver ailments and the third is useful in diagnosing myocardial infarction. Commercially available reagent tablets were dissolved in triply distilled water and added to silicone rubber pads, as previously described. In use, 50 microliters of distilled water were added to the pads to reconstitute the reagents. The well-mixed solution was spread evenly over the pad and inserted into the fluorometer to measure any background fluorescence (fluorescence not due to the enzyme reaction). Of the sample 10 microliters were then added to the pad and mixed with the reagents. The change in fluorescence was recorded immediately in the HBD reaction;

SGOT and SGPT require a 3-minute incubation before useful data are obtained. Again, calibration curves are used to relate the change in fluorescence to the activity of the enzyme measured.

Automated Methods of Enzymic Analysis

In recent years, devices have appeared on the market which automatically add and mix reagents with samples, deliver the reaction mixture to some detector, and, quite frequently, provide the results on a printout, a graph, or a digital display. Such methods are not reagentless, but, in cases where very many samples of the same kind are to be analyzed, they may save a great deal of technician time. It should not be surprising to learn that a number of enzymic assays have been automated.

Some very significant automated methods have been reported by a group of Scottish scientists, and we shall outline a small part of their work. In their earlier studies, polystyrene was used as an enzyme support. Subsequently, they devised methods for bonding enzymes to nylon and this nylon-enzyme adduct was used in automated analysis. As examples of their method, we shall consider automated analyses of ethyl alcohol and pyruvic acid.

The enzymes, alcohol dehydrogenase (ADH) and lactate dehydrogenase (LDH), were used as part of the automatic analyzer. The substrates, ethyl alcohol for ADH and pyruvic acid for LDH, were contained in a reservoir, from which measured samples could be withdrawn automatically. The substrate was aerated after flow was started. Cofactors, NAD for ADH and NADH for LDH, were pumped through debubblers and thoroughly mixed with the substrates in mixing chambers. The mixture was then passed through the enzyme-containing nylon tubes, where the reaction occurs. The reaction of ADH converts NAD to NADH, while the LDH reaction works in the opposite direction. The reactions are illustrated in Equations 9–17 and 9–18. The conversion is

Equation 9–17

Ethyl alcohol and NAD $\xrightarrow{\text{Alcohol dehydrogenase}}$ Acetaldehyde and NADH

Equation 9–18

Pyruvate and NADH $\xrightarrow{\text{Lactate dehydrogenase}}$ Lactic acid and NAD detected at 340 nanometers in a spectrophotometer with flow-through cells. The system is automatically washed between assays and 30 samples can be assayed per hour. Calibration curves are prepared by graphing the reading at 340 nanometers as a function of known substrate concentrations. Unknown solutions are determined by comparing the reading they give at 340 nanometers with the calibration curve.

The enzyme tubes were filled with solutions of substrate and stored at 4° C when not in use. They were used over a period of 20 days without loss of activity and, during this time, each tube was used for at least 1,000 assays. Solutions of the enzymes, similarly stored, lost at least 90% of their activity in the same time.

A commercially available automatic analysis instrument, based on immobilized enzymes, has been described by scientists at Leeds and Northrup Company at North Wales, Pennsylvania. Their device uses glucose oxidase bonded to glass particles for measuring glucose in solution. By inserting a column of lactase before the glucose oxidase, lactose can be measured; columns containing invertase and mutarotase preceding the glucose oxidase column yield a system for measuring sucrose concentrations.

The instrument is used in an end-point (as opposed to rate) type of assay. It can be used to measure individual samples or for continuous monitoring of flowing streams. To determine the sugar content of a flowing system, a probe containing a semipermeable membrane is used. The membrane permits the small substrate molecules to enter freely, but holds out large molecules, which are not important in the assay and which might otherwise foul the system.

A peristaltic pump propels the sample through columns of glass-supported enzyme, where conversion is completed. The temperature of the enzyme column is automatically controlled. The detector is sensitive to the hydrogen peroxide produced in the reaction of glucose with glucose oxidase (see Equation 9–3). The other sugars are measured by converting them to glucose, fructose, and galactose and measuring the peroxide formed when this glucose is reacted with glucose oxidase and oxygen. The readout used is an illuminated digital display. It can be programmed to read sample concentrations in any desired units.

The enzyme columns, used in continuous monitoring, retain activity for 3 to 5 weeks. When activity is lost, the column is simply removed and replaced.

The instrument has been designed for industrial purposes; thus, automatic enzyme analysis is possible not only in the clinical laboratory, but also in the industrial laboratory. It can be seen that, with the advent of immobilized enzymes, the role of enzymes in chemical analysis can be described by the words of a country-western song—"It ain't a big thing, but it's growing."

Chapter Ten

Enzymes in Medicine[1]

THE USE OF enzymes for therapeutic purposes has had a long and controversial history. Enzyme therapy was extolled and reviled around the turn of the present century, just as it is today. It seems that there is a small, but enthusiastic, group of physicians who use enzymes regularly and confidently, and a larger group which is skeptical and avoids the use of these therapeutic tools. Enthusiasm may be based more on pragmatic grounds than scientific ones. Enzyme therapy is used because it works. Critics of enzyme therapy consider the abundant reports of benefits of such therapy to be "testimonials" rather than scientific studies. But carefully controlled "double-blind" studies are being reported with increasing frequency, giving solid scientific support to enzyme therapy.

A conference on "Proteolytic Enzymes and Clinical Applications" was held by the Section of Biology of the New York Academy of

---

[1] It is useful to learn a few important phrases, words and parts of words used in the medical section. A "blind" study is one in which the patient does not know whether he is receiving a drug or a placebo ("sugar pill"); in a "double-blind" study, neither the patient nor the therapist knows. Hyper means high; hypo means low. Uria means in the urine; -emia means in the blood. Thus, hyperglycemia means high glucose in the blood. An acute illness is one which has a sudden beginning, a short course, and severe symptoms; a chronic illness is one of long duration.

Science in 1956. More than a dozen clinicians told of impressive results achieved by use of enzyme therapy. At the conclusion of the conference, Dr. O. H. Stuteville of the Northwestern University School of Medicine in Chicago said, "For my part, I am not convinced of everything that has been claimed for the proteolytic enzymes, but I feel that they are of definite value. I have seen some phenomenal results from this therapy. However, I feel that we must prove it to ourselves as well as to those who doubt it more than we do."

In the 20 years since that conference, many more reports of successful enzyme therapy have appeared and many more physicians have been won over. Still, it seems appropriate for us to share the conservative approach inherent in Dr. Stuteville's statement, as we outline some of the uses of enzymes in therapy. With this in mind, some of the criticisms of enzymes as therapeutic agents should be mentioned. Probably the three most persistent criticisms are:

1. It is not known how enzymes exert their therapeutic effect.
2. Although there are many reports of beneficial uses of enzymes in medicine, there are few scientifically controlled studies.
3. Enzymes are foreign proteins introduced into the body. As such, they are capable of producing reactions which can range from a mild case of hives to severe, life-threatening anaphylactic shock.

All of these criticisms are valid, but all can be answered. It is not known how enzymes work, but it is not known how aspirin reduces fever, or how gold preparations help some arthritics. Many therapeutic agents work by unknown mechanisms; their use persists because they are effective.

The number of controlled studies is not as great as one would hope, but the number is increasing gradually. All of the controlled studies which have been published seem to support the therapeutic effectiveness of enzymes.

The danger of a severe reaction to a foreign protein is very real and patients should be watched carefully for such reactions. Yet, many therapists have reported highly successful treatments with enzymes with no adverse effects. It should be remembered that insulin, which keeps so many diabetics alive, is also a foreign protein.

Results of enzyme therapy reported before World War II can, for all practical purposes, be discounted. The study of enzymes was in a very rudimentary stage and enzyme purification methods were meager.

Thus, it is impossible to know whether any effect, beneficial or harmful, should be attributed to the enzyme or to some unknown impurity.

Just as the first half of the twentieth century was an age of exponential growth in the accumulation of knowledge in physics, so the second half of this century is bringing rapid understanding of biochemical processes. The complicated structures of many enzymes have been mapped and enzyme synthesis in the laboratory has been achieved. Enzyme purification has reached a sophisticated level with the discovery of physical techniques, such as gel-permeation chromatography and chemical methods, including affinity chromatography (see Chapter 15). Pure, sterile enzyme preparations are commercially available for clinical use. In this, and the next three chapters, we shall consider some of these uses.

Inflammation and Enzyme Therapy

The normal functioning of living cells and tissues can be disturbed by a variety of means. These include chemical means, as by the ingestion of toxic substances, physical means, as an athletic injury, and biological means, such as by germs and viruses. At the moment of the disturbance, a series of defensive measures start to counteract the disturbance. These measures are collectively called the inflammation reaction.

This normal protective response can proceed more vehemently than necessary. In trying to create an environment which is hostile to an invading microorganism, the tissues often create conditions unsuitable for their own survival. The permeability of blood vessels in the area of the trauma increases and plasma exudes into the injured area. In this unnatural environment, the plasma proteins become partially denatured making the fluid viscous. The permeability of the blood vessels then decreases and the blood clotting mechanism is "turned on" in this area. All channels for fluid exchange are blocked. The accumulation of excess fluid in the disturbed area is called edema. It is one of the primary symptoms of inflammation and is partially responsible for the clinically observed secondary symptoms: heat, redness, swelling, and pain. It has been said that any inflammation which is clinically observable should be treated.

A number of proteolytic enzymes have been used in the treatment of inflammation. These include trypsin and chymotrypsin from the pancreas of cattle, papain from the papaya tree, bromelin from pineapple stems and streptokinase-streptodornase from streptococcus (bacteria).

Dr. Irving Innerfield of the Kings County Medical Center and Long Island University in New York has provided abundant evidence for the efficacy of enzymes in treating inflammation. He has written numerous scientific papers describing his many successful therapeutic uses of enzymes and, in 1960, published a book documenting the historical, experimental, and clinical evidence for the rational use of enzymes as therapeutic and diagnostic agents.

Innerfield studied the clinical effects of trypsin in 538 patients and noted dramatic and rapid suppression and reversal of all signs and symptoms of acute inflammation in 94.3% of the patients. The fluid of edema was resorbed and heat, redness, swelling, and pain were relieved.

Dr. A. Lee Lichtman used trypsin to treat professional boxers, who "did their thing" in Madison Square Garden. He has found that an injection of trypsin in the buttocks immediately after injury has made black eyes and bruises subside in 1 to 3 days, rather than the normal 10 to 14 days. Early treatment was stressed, since trypsin had little effect once discoloration was seen.

Cetrulo injected trypsin into the vein of a gunshot victim. Edema was cured, the wound was clean, drainage stopped, the exudate was minimized, and the edges of the wound appeared fresh and healthy.

Trypsin is often used in combination with chymotrypsin in the treatment of inflammation. This combination seems likely to be exploited more fully since recent, carefully controlled studies have demonstrated significant improvements in postsurgical inflammation and in athletic injuries.

Papain is available commercially in a green peppermint flavored tablet and has been in general use since 1963. Controlled, double-blind studies have supported the use of the enzyme in treating obstetrical (childbirth-related) inflammation and edema, as well as the inflammations due to dental surgery and to obstruction of the ureter (tube leading from the kidneys to the bladder). Dr. Henry Holt of Wayne State University used papain in a 2-year study of the treatment of athletic injuries and concluded that mild injuries, most frequently bruises, showed a high frequency of favorable response to papain as an anti-inflammatory therapy.

The medical literature contains numerous reports of beneficial effects of trypsin, chymotrypsin, and papain in treating inflammation. These few examples are given only as illustrations. Bromelin, streptokinase, and streptodornase have also been used as anti-inflammatory agents; all three are useful in other areas of therapy as well. Thus, bromelin is used in correcting digestive disorders and the streptokinase-streptodornase mixture is used as a thrombolytic agent.

ENZYMES IN THROMBOLYTIC THERAPY

We have seen that the inflammation reaction is an example of a defense system which can overreact and require therapeutic intervention. The clotting of blood provides another instance of this. It is now known that, in normal tissues, there is a constant dynamic equilibrium between blood coagulation (clotting) and fibrinolysis (the process of dissolving the clotted blood). The maintenance of proper balance in this equilibrium is extremely important. If fibrinolysis is increased by a pathological cause, a predisposal to excessive bleeding results. On the other hand, if fibrinolysis is weakened so that clot formation is favored, conditions occur that are called thromboses (clots formed in and remaining in blood vessels) and embolisms (sudden blockages of blood vessels caused by circulating fragments of clots). These can be life threatening. In chronic cases, cholesterol and other fatty materials may aggregate around clotted deposits in blood vessels. When this pathological condition is well established, it is called atherosclerosis—hardening of the arteries.

Although there are chemical anticoagulants (such as heparin) available as well as corrective surgical techniques, acute thromboses and embolisms (which are lumped together as acute thromboembolic vascular diseases) are still the largest single cause of death and disease in the middle-aged and elderly populations of the Western World.

The mechanism used by the body for controlling the equilibrium between clot formation and dissolution is a complex one involving a series of enzymes, proenzymes, activators, and proactivators. A brief outline of the highlights of the process will be helpful in understanding the therapeutic method. In forming a clot, the plasma protein, fibrinogen, is converted to insoluble fibrin by the enzyme thrombin. Fibrin forms the clot. The enzyme plasmin, which dissolves the clot, exists in the blood as the proenzyme plasminogen. Activators convert plasminogen to plasmin for dissolving the clot.

Just as nature uses enzymes in maintaining this crucial balance, so is man learning how to use enzymes to restore the balance once it is lost. Clinical studies have shown that the best approach to therapeutic thrombolysis (dissolution of clots) is an intravenous injection of an enzyme capable of converting plasminogen to plasmin—the enzyme which dissolves the clot. This type of therapy is known as thrombolytic (thrombus- or clot-splitting) or fibrinolytic (fibrin-splitting) therapy. The enzymes most frequently used for this are streptokinase from bacteria and urokinase from human urine. Three "new" thrombolytic enzymes which have seen some therapeutic use in the last 20 years are arvin (from a Malayan pit viper), reptilase (from a South American snake), and brinase (from the mold Aspergillus oryzae).

Fibrinolytic therapy has been used successfully in the treatment of occlusions (blockages) of veins and arteries. It can be a lifesaving treatment in cases involving pulmonary embolisms and myocardial infarctions (blockages of vessels in lungs and heart, respectively). A pulmonary embolism is usually composed of a recently formed clot. Its first effects produce strain on the right side of the heart—the side which pumps blood to the lungs.

Several groups of physicians have reported successful lysis of pulmonary embolisms by use of enzymes. These agents are most effective in massive pulmonary embolisms involving large blood vessels, since clots in these vessels are more accessible to circulating enzymes than are those in smaller arteries of the lungs. Surgical removal of a pulmonary embolism is a dangerous procedure, so it seems likely that the use of thrombolytic therapy will increase in patients afflicted with this condition.

A group of clinicians from various hospitals, reporting together as the "European Working Party" in 1971, described a controlled experiment which confirmed the superiority of streptokinase over heparin (a chemical anticoagulant) in reducing deaths due to acute myocardial infarction. Both the total number of deaths and the number of deaths within 24 hours of being stricken were significantly lower in the enzyme-treated patients. There was also a significant decrease in the number of infarcts reformed and in the number of deaths due to heart failure in the enzyme-treated group when compared with the heparin-treated patients.

A danger in fibrinolytic therapy is the possibility of a clot reforming. After a clot is dissolved, the tissue in the area remains damaged and the likelihood of new clot formation at that spot is high. Therefore, fibrinolytic therapy is followed up with anticoagulants, such as heparin.

Major problems associated with streptokinase therapy are fever, a tendency to bleeding, antigenicity (as in any foreign protein), and the difficulty of determining the proper dose. Bleeding seems to be the most serious of these; it is also a problem when anticoagulants are used alone.

Urokinase (formed in the kidneys and obtained from human urine) seems to be safer than streptokinase, but, because of the difficulty and expense in producing it, clinical exploitation has been limited. (2,300 liters of urine yield 29 milligrams of highly purified urokinase. A milligram is one-thousandth of a gram.)

Arvin and reptilase have been used less frequently but are possible successors to heparin as anticoagulants. Controlled studies of these enzymes and of brinase are needed before they are widely accepted for therapeutic uses.

ENZYMES VERSUS VIRUSES

Viruses are large molecules composed mostly of nucleic acid, usually having a coating of protein. They are not considered living since they have no known metabolism and can multiply only after invading a host cell. The normal activities of the host cell are lost after viral infection; it seems to devote all of its energy to reproducing viral substances. Although relatively few viruses are pathogenic, those that are cause a large variety of diseases in man and animals. Many scientists believe that cancer is caused by viruses, and it has been found that a number of types of malignancies can be induced in laboratory animals by viral infection.

The protein coating is believed to be important in the attraction and attachment of the virus to the host cell and in the actual invasion of the virus into the cell. It is the protein coating that is believed to be the most logical point at which to attack the virus molecule. Thus, in this way enzymes have been used to a limited extent in treating viral diseases in animals and man. Wild and Brown reported in 1967 that trypsin inactivates the virus for hoof-and-mouth disease, completely eliminating its infectivity. Cleeland, in 1963, achieved similar results with trypsin and influenza virus. Miller (1956) reported successful treatment of mumps with streptokinase. Gotz has used proteolytic enzymes against the viruses herpes zoster and herpes simplex (fever blister virus) in man.

There are few effective treatments for viral diseases. It seems likely that enzyme therapy and prophylaxis (prevention of disease) will provide significant protection against viral infections as knowledge of their effectiveness accumulates and spreads.

ENZYMIC TREATMENT OF WOUNDS, BURNS, SCARS, AND ULCERS

Collagen is the main constituent of skin, tendon, and cartilage and is the protein component of teeth and bones. About one-third of the total protein in mammals is collagen. The collagen molecule consists of a triple helix of polypeptide chains with each chain containing approximately 1,000 amino acid residues. The amino acid composition of collagen is unusual in that it has a very high percentage of glycine, proline, and alanine and it seems to be the only mammalian protein containing hydroxyproline. The concentration of this amino acid increases with age. Sulfur-containing amino acids and tryptophan are missing. Collagen is resistant to the usual proteolytic enzymes, but it can be decomposed by specific enzymes known as collagenases. These enzymes have been found useful in the debridement process of removing devitalized tissue from wounds, burns, scars, and ulcers.

The presence of surface debris and dead tissue in and around a wound can retard tissue repair. Such materials are often anchored to the surface of the wound by strands of collagen. These strands must be broken for debridement to occur. This process is necessary for prevention of infection.

Collagenase has been found effective in the debridement of surgical wounds and it reduces the extent of scar tissue. Probably its most widespread uses to date have been in the treatment of burns and ulcers.

A third-degree burn contains collagen in varying stages of denaturation. That in the charred center is completely denatured; fibers around the rim are less denatured; and the collagen of the surrounding skin is undenatured. Completely denatured collagen can be decomposed by proteolytic enzymes, but the less denatured collagen can be decomposed only by collagenase.

Dr. E. L. Howes and co-workers reported a series of successful animal experiments involving burn therapy with collagenase. This was followed by clinical trials in which small burns were treated with the enzyme in a sterile ointment. The first three cases involved burns of the forearm, which responded very well to the therapy. The slough came off within 24 hours with no complications.

These successes were followed by a number of failures, but the failures provided extremely useful information. Howes and his group discovered that the failures were not due to any inherent deficiency in the enzyme therapy, but rather to the fact that antiseptics and cleaning solutions, which were used in first aid and in cleaning the burned area, caused cross-linking of the surface of the burn. This cross-linking gives collagen a leatherlike quality and makes it impervious to enzymic attack. The cross-linking agent was identified as hexachlorophene. The same effect can be caused by tannic acid, chromic acid, quinones, some oils, silver nitrate, and other chemicals capable of cross-linking proteins. Dr. Howes has pleaded that non-cross-linking antiseptics and cleaning agents be used in first aid and in early treatment of burns, so that collagenase therapy can be used at the proper time, which is normally 3 or 4 days after the burn.

A. O. Varma and co-workers found collagenase to be an effective debriding agent in dermal (skin) ulcers and decubitus ulcers (bedsores) in a double-blind study of 2 groups of 10 patients. They found that the enzyme liquefies the dead debris of a wound without damage to healing tissues and provides a clean base for subsequent transplants. No adverse clinical effects were seen except for redness of surrounding skin in one patient. This condition was corrected. They concluded that collagenase is a useful drug in the management of chronic ulcers contaminated with dead tissue and of decubitus ulcers. They felt that, because of its unique ability to hydrolyze undenatured collagen, it may be the drug of choice for wound debridement. They repeated the caveat that the enzyme is inactivated by a number of materials which should not be used when collagenase therapy is indicated.

Because materials other than collagen are involved in burns and wounds, enzymes other than collagenase have been used as debriding agents. For instance, a mixture of streptokinase and streptodornase was found to be effective in decomposing susceptible portions of wound exudates and sloughs in a variety of surgical lesions. Pus is made up largely of dead cells and cell debris. In such material, there is abundant deoxyribonucleic acid (DNA); this is the natural substrate for streptodornase. Streptokinase is an effective agent for hastening the breakup of blood clots, which may contaminate a wound.

ENZYMES IN AGING

As a person ages, his enzyme supply decreases in amount and in activity. These decreases are probably at least partially responsible for the development of the characteristic symptoms of aging, and for premature aging. A trivial example of this is provided by the fact that the greying of hair has been attributed to a lack of tyrosinase, or a loss in activity of the enzyme, with advancing age.

In the section on thrombolytic therapy, we saw that plasmin, and the proenzyme plasminogen were important in maintaining the equilibrium between clotting and clot lysis. Astrup has found that the synthesis of plasminogen and of plasminogen activators are reduced during aging. This, of course, favors the formation of fibrin deposits in the blood vessels. Cholesterol and other fatty materials tend to become associated with the fibrin and this can lead to hardening of the arteries—a condition which leads to heart attacks and decreases the functioning of the brain, kidneys, eyes, and ears.

Drs. Max Wolf and Karl Ransberger, who are pioneers in the therapeutic use of proteolytic enzymes, have used a mixture of proteolytic enzymes, often in combination with vitamin E, to treat senior citizens. Several hundred patients have been helped by the therapy. Among the benefits already observed are an increase in the clot lysing activity in the blood and a reduction of serum lipids (including cholesterol) to normal.

ENZYMES AS DIGESTIVE AIDS

Another condition, common in older people, although by no means limited to them, is an impaired ability to digest foods. The problem is more severe in older people because all digestive secretions, including amylase, pepsin, hydrochloric acid, bile, and the pancreatic enzymes, decrease with age.

The use of digestive enzymes to supplement the body's natural supply seems to have provided the first successful example of enzyme replacement therapy. A wide range of digestive enzymes are available to the clinician for treating problems of digestion.

A number of these preparations have enteric coatings to protect the enzymes from attack by pepsin and hydrochloric acid in the stomach. Some physicians have reported that enteric coating is unnecessary. If the enzymes are taken when the stomach is full, the pepsin and hydrochloric acid are apparently diluted to the point that they do not deactivate the therapeutic proteins. In addition, if the patients are deficient in gastric (stomach) juices, there is little likelihood of enzyme deactivation in the stomach.

ENZYMES IN CATARACT SURGERY

Cataract operations involve the removal of the eye's lens that has become opaque, usually due to aging. The lens must be detached from the zonules (ligaments) which hold it in place. This tearing can damage other parts of the eye and delay healing.

A Spanish surgeon noted that the use of proteolytic enzymes could reduce the occurrence of such problems. The surgeon makes an incision in the cornea and exposes the zonules to a dilute solution of chymotrypsin for 2 to 4 minutes. The protein-dissolving enzyme selectively breaks down the zonules permitting the lens to be removed easily. The enzyme does not damage other parts of the eye. After the lens is removed, the patient usually regains vision by use of eyeglasses.

PENICILLINASE

Many years ago, a professional comedian noted that medical science had advanced tremendously because a cure had been found for penicillin. It may have sounded funny at the time, but it is substantially correct. Severe reactions to penicillin are a common occurrence. An enzyme which converts penicillin to penicilloic acid (nonantibiotic and nonantigenic) was discovered in 1940, used experimentally on human subjects in 1956, and released for general use in 1958. Dr. Murray Zimmerman of the University of Southern California Medical School called penicillinase the most specific enzyme used in clinical medicine and the best understood. It has but one known effect—the hydrolysis of penicillin.

The enzyme has favorably modified the course of penicillin reactions in over 80% of cases treated from simple urticarias (hives) to severe, life-threatening anaphylactic shock and shocklike syndromes. The enzyme destroys the causative factor (penicillin) and symptoms disappear.

Side reactions are typical of responses to foreign proteins: pain and hardness at the site of injection, fever, hives, and, rarely, anaphylactic reactions. Although the enzyme itself has side reactions, it is more successful in treating penicillin reactions than other available drugs. Repeated penicillinase treatment in man will not be common, since penicillin-sensitive patients will avoid further exposure to this antibiotic. Antipenicillinase has long been available to counteract the enzyme in an emergency.

Penicillinase therapy does not correct the harm done in the antigen-antibody reaction. It merely removes the irritant (penicillin, which is the antigen). It may be compared to a burn. We remove the agent causing the burn, but the effects of the burn linger.

None of this should be interpreted as a denigration of penicillin; we know that no drug is without side effects. For example, deaths have been caused by aspirin. Penicillin is still a highly useful antibiotic. Only a few antibiotics are bactericidal. Of these, only penicillin is nontoxic.

Dr. Alexander Fleming, the British bacteriologist who discovered penicillin, is also credited with the discovery of lysozyme, an enzyme which has bactericidal properties. Legend has it that Dr. Fleming accidentally sneezed into one of his bacterial culture dishes and watched in utter amazament as the bacteria dissolved. As a result of this observation, he discovered lysozyme in human saliva.

Lysozyme Ampicillinate

It is noteworthy that both penicillin and lysozyme are active against the bacterial cell wall. Penicillin inhibits the synthesis of teichoic acid—one of the components of the cell wall—while lysozyme disrupts the cell wall by hydrolyzing components known as mucopeptides. Both agents have additional properties which make them useful in fighting microorganisms. Fleming noted that penicillin stimulates phagocytosis—a process by which the body engulfs and removes foreign particles. Lysozyme inhibits respiration of bacteria and, in high enough concentrations, blocks respiration completely. Thus, it can kill microorganisms without breaking the cell wall. Lysozyme is found in biological fluids and also inside cells. It is believed that lysozyme, in the cells responsible for phagocytosis, acts to kill and decompose the bacteria which the cells have engulfed.

Lysozyme is also active against viruses and has been effective in treating viral infections. The antivirus activity is due to the amino acid composition of lysozyme. It has a high percentage of basic amino acids which make the molecule as a whole basic. Most viruses seem to be acidic (since they are composed primarily of nucleic acids). Scientists have suggested that the acidic virus combines with basic lysozyme forming an insoluble complex which has neither the cell-destroying nor infectious action of the native virus. The antiviral therapeutic activity of lysozyme has been confirmed by successful treatment of viral skin disease, influenza, and viral hepatitis. The enzyme may even be useful in preventing the two latter diseases.

Still another therapeutic property of lysozyme is its anti-inflammatory activity. This has been shown in the treatment of patients suffering from acute abdominal infections. Patients treated with lysozyme and penicillin after surgery lost the inflammatory symptoms more quickly than those treated with antibiotics alone.

Italian workers have noted that lysozyme and penicillin are more effective when used together than either is alone. Scientists call this "synergistic action." The synergism is believed to be due to the breakdown of the cell wall by the enzyme permitting the antibiotic to act to a greater extent. Russian physicians used lysozyme and penicillin to treat patients who were infected with penicillin-resistant staphylococci. The success of the therapy confirmed the synergistic action of the drugs.

The synergism is an important discovery because it makes it possible to produce good therapeutic effects with less toxic antibiotics. Pellegrini and Vertova of the Alexander Fleming Research Institute in Milan have combined lysozyme and penicillin forming a salt called lysozyme ampicillinate. The salt is expected to display all of the advantages of the combined drugs.

HYALURONIDASE

Hyaluronidase is an enzyme that is found in many strains of bacteria, in bee, snake, and spider venoms, and in mammalian testicles. It first became clinically useful as a "spreading factor" because it increases dispersion of injected fluids. The natural substrate for this enzyme is hyaluronic acid—a material found in all connective tissues of man and animals (see Figure 10–1).

Figure 10–1   Hyaluronic acid.

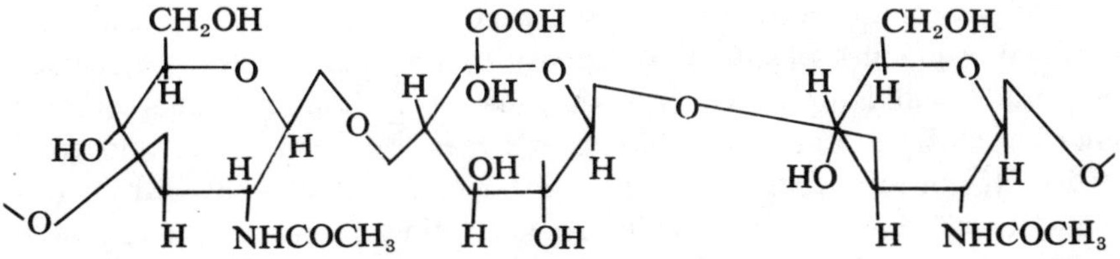

The function of the material seems to be to bind water in interstitial spaces and to hold cells together in a jellylike matrix. It may also provide the fluids present in joints with shock-absorbing properties.

By reducing the viscosity of the gellike matrix, hyaluronidase temporarily facilitates the dispersion of fluids, electrolytes, and metabolites. Its action is purely local and it has no effect on the spread of local infections, providing it is not injected into the infected area.

Hyaluronidase has been found of value for local anesthesia in surgery, dentistry, and opthalmology by increasing the area of anesthesia. However, it decreases the duration of its effectiveness.

Natural hyaluronidase inhibitors are found in the plasma, and salicylates (such as aspirin) inhibit the enzyme.

Maroko reported in 1972 that intravenous injection of the enzyme caused a decrease in myocardial edema (fluid in the heart muscle) in patients with acute myocardial infarctions. This can be a life-saving technique, since highly lethal cardiogenic shock is associated with large infarcts. This potential life-saving action of hyaluronidase is attributed to its effect on the transport of energy-producing nutrients from the bloodstream to the myocardial cells.

Hyaluronidase depolymerizes hyaluronic acid, thereby easing the transport of a variety of substances through the interstitial spaces. It increases capillary permeability. This may be particularly important in the presence of coronary occlusion, in which nutrients have to be transported through longer extravascular pathways than when the coronary arteries are open. Whatever its mechanism, hyaluronidase protects the myocardium from undergoing excessive ischemic injury and consequent necrosis following coronary occlusion. This action is without apparent deleterious side effects and suggests that this enzyme may be of value in the treatment of patients with an impending myocardial infarction, or early after a coronary occlusion.

Some Dental Uses of Enzymes

One of the most important aspects of oral hygiene is the daily removal of plaque—a soft sticky deposit that accumulates on the surface of teeth. It builds up on clean teeth in about 24 hours. Plaque contains microorganisms, protein (including enzymes), carbohydrate, and food debris. If not removed, it turns into tartar or calculus—the hard deposit that dentists remove mechanically. As layer piles on layer, it gives rise to pathological conditions of the teeth and soft tissues of the mouth. It can cause teeth to separate from the gums.

A good deal of research has gone into plaque control—some of it involving enzymes. Oral hygiene preparations containing combinations of proteases, amylases, lipases, and cellulases in various proportions have been compared. The consensus seems to be that those preparations containing a high percentage of proteolytic enzymes are most effective against plaque.

It is known that streptococci contribute to plaque formation by forming carbohydrate polymers known as dextrans. These polymers serve as a part of the matrix of some plaques. Dextranases have been found to loosen and disperse artificial plaques formed by streptococci. Although the bacterial flora of the mouth do not always contain streptococci or other dextran-forming microorganisms, it seems likely that a dextranase should be included in enzyme-containing oral hygiene products.

The instability of enzymes causes difficulty in obtaining storable, usable, enzyme-containing dental hygiene materials. An interesting approach to solving this problem was described by B. S. Wildi and T. L. Westman in a patent issued to Monsanto Corporation.

They bonded various enzymes to polymers such as ethylene-maleic anhydride copolymer (EMA). The enzyme-polymer adduct can be made in soluble and insoluble forms. The adducts are much more stable than free enzymes, and, when the enzyme is a dextranase, an added benefit is realized. The optimum pH for free dextranase is in the acidic range, around 4 to 5, while the pH of the mouth is closer to neutrality—between 6 and 8. The EMA-dextranase adduct has a pH optimum between 6 and 8, making it most active under conditions found in the mouth. The material is said to be useful in dental hygiene mouthwashes, gargles, gums, and toothpastes.

Plaque and stain removal are not the only uses of enzymes in dentistry. Dr. G. D. Magnes has reported the results of a double-blind study he carried out to determine if papain was effective in treating edema and other complications of dental surgery. He found significant benefits in the reduction of edema, inflammation, and pain. No side effects were seen. Dr. Magnes concluded that papain is a safe and effective adjunctive therapy in the control and reduction of edema, inflammation, and pain associated with trauma resulting from oral surgery.

URATE OXIDASE IN TREATMENT OF GOUT

Purines are compounds which occur in high levels in nucleic acids—the carriers of genetic information in the cell. They are also found in foods comprising a normal diet, especially meats. Purines undergo metabolic changes to uric acid and then to allantoin in normal, healthy people. Structures of purine, uric acid, and allantoin are shown in Figure 10–2.

Figure 10–2

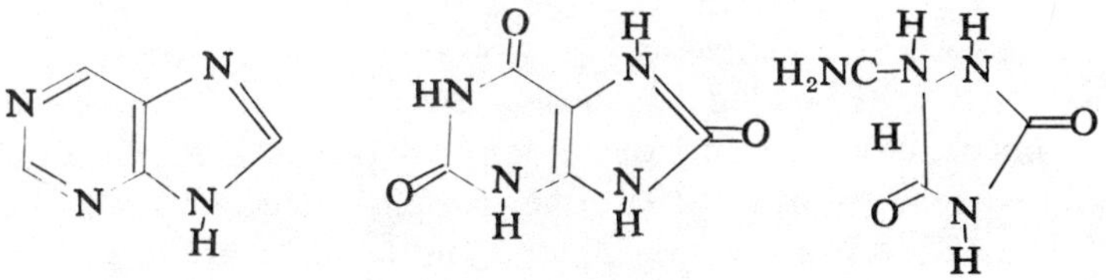

Purine   Uric acid   Allantoin

Allantoin is slightly more soluble than uric acid and is cleared rapidly and completely by the kidneys, resulting in rapid removal from the blood. Those afflicted with gout apparently cannot convert uric acid to allantoin. Sodium urate crystallizes in the cartilage producing extreme sensitivity of the joints, which characterizes gout. This painful and sometimes debilitating disease is probably due to an "inborn error of metabolism" (see Chapter 13).

The conversion of uric acid to allantoin by urate oxidase was demonstrated in animals in 1949. Injections of urate oxidase, obtained from an animal source, were used to treat gout patients in 1957. After such injections, a patient suffering from long-term gouty arthritis was able to convert uric acid to allantoin. The activity lasted only a few hours and frequent daily injectons were needed. Antibody formation ultimately forced an end to the therapy.

A French group isolated a urate oxidase from Aspergillus flavus (a fungus). The enzyme was purified and found to have a high level of activity in animals. Its antigenic potential was moderate and toxicity was negligible. It could be produced in industrial quantities.

In 1968, French physicians reported the use of this enzyme in a study involving 61 hospitalized patients, all of whom received a stabilized purine diet. They were divided into three groups: normal, hyperuricemic, and patients with gout. Intravenous or intramuscular injections of the enzyme were given at 24-hour intervals. Three typical gouty attacks were interrupted 24 hours after therapy began. Uric acid elimination was increased fourfold over pretreatment levels. Softening of the crystal deposits in gout patients indicated that the deposits were being depleted. The physicians concluded that "The only practical medication capable of destroying uric acid 'in situ' (in the normal place) is fungal urate oxidase; its use in the management of gout and hyperuricemia is indisputable." Unfortunately, antigenicity problems were ultimately also seen in this treatment. These problems limit the usefulness of the therapy, but are not dangerous.

In 1975, J. C. Venter and co-workers at the University of California at San Diego bonded bacterial urate oxidase to glass beads. In one experiment, blood taken from a dog was passed through a column containing the glass-supported enzyme. Results were equivocal because of low uric acid levels in the dog's blood and because of the presence of allantoin in the blood. Human blood with added uric acid was then continuously circulated through the column. After 4 hours at 37° C, all uric acid was converted to allantoin.

A dog was fed a special diet for a week to increase the uric acid level in the blood. When the uric acid reached a relatively high level and stabilized at that level, an extracorporeal shunt (a device by which blood is taken from the body, treated externally, and returned to the body) was fitted to the dog. The dog's blood was passed through a column of glass-supported uric oxidase and back to the body. A substantial decrease in serum and urinary uric acid was seen, along with a corresponding increase in urinary allantoin levels. The postoperative period was routine with no detectable ill-effects resulting from the treatment. There was little detectable loss of enzymic activity in the column resulting from this use.

Antibodies which were able to cause complete inhibition of native uricase failed to inhibit the supported enzyme.

The authors believe that a combination of this treatment with kidney dialysis may be useful in preventing the hyperuricemia resulting from such dialysis. They suggest further laboratory studies before clinical use is attempted.

ENZYMES AND THE ARTIFICIAL KIDNEY

Every year, many thousands of people die from kidney disease in this country. An estimated 6,000 to 10,000 persons could probably be saved by kidney transplantation or hemodialysis.

As now practiced, hemodialysis therapy requires the patient to be tied to his artificial kidney two or three times a week for approximately 8 hours each time. During therapy, blood flows through a plastic cannula (a small tube) inserted into an artery in arm or leg, through the artificial kidney, and back to the body through another cannula inserted into a vein.

While in the artificial kidney, the blood is separated from an aqueous dialysis solution by a membrane usually made of cellulose or some derivative of cellulose. Small molecules and ions can easily flow through the membrane, but blood cells and large molecules, such as proteins, cannot cross the membrane. Waste products, such as urea and creatinine, pass from the blood through the membrane into the dialysis solution. This process can continue until the concentration of the waste product in the dialysis solution is equal to its concentration in the blood. At this point, equilibrium is reached and no further clearance of the waste product can be achieved unless the used dialysis solution is replaced with fresh solution.

Many of the small molecules and ions in the blood are essential and removing them in the dialysis treatment would be harmful. To avoid this problem, the dialysis solution is prepared in such a way that it contains these small molecules and ions in concentrations approximately equal to their concentration in the blood. Thus, these materials do not diffuse out of the blood into the dialysis fluid. (Stated more accurately, the materials diffuse back and forth through the membrane maintaining approximately equal concentrations on both sides of the membrane.) Usually, an extra pressure is exerted on the blood or a vacuum is applied on the dialysis fluid side of the membrane to help remove water from the blood.

While the artificial kidney is saving a goodly number of lives each year, it has a number of shortcomings. Only 1.5% of the people needing it are being helped by the artificial kidney or transplants. Additional drawbacks include high capital and operating costs, the need for large volumes of dialysis fluid, and the requirement of visiting the device two or three times a week. This unnatural method of eliminating wastes results in a large buildup of metabolic products between treatments and a rapid removal during treatment. This is neither comfortable nor desirable.

It requires 100 to 300 liters of dialysis fluid to treat a patient with the artificial kidney. If metabolic wastes could be removed from the fluid, these large quantities would not be necessary. A system could then be designed to permit more portable and more frequent (and, presumably, more generally available) treatment.

As a first approach to the removal of waste products from the dialysis fluid, adsorption was tried. This method is quite useful for the removal of waste materials such as creatinine, uric acid, and phenols, but it has not proved satisfactory in removing the most abundant metabolic waste found in the urine—urea.

Dr. Thomas M. S. Chang of McGill University in Montreal has proposed the use of "artificial cells" to solve this problem. The artificial cells, in this instance, consist of microencapsulated urease and a microencapsulated adsorbent for ammonia. Urease decomposes urea, as shown in Equation 10–1.

Equation 10–1

$$H_2NCONH_2 \text{ and } H_2O \xrightarrow{\text{Urease}} CO_2 \text{ and } 2NH_3$$

Urea and Water   Carbon dioxide and Ammonia

Carbon dioxide is neutralized in the blood. Ammonia is much more toxic than urea, and it must be removed from the circulation. An efficient adsorbent for ammonia is needed for the artificial cell approach to removing urea.

Urease-containing artificial cells could be used in the dialysis fluid, or in a separate section of the artificial kidney. Still another mode of use is possible: the patient could ingest the artificial cells. The enzyme would hydrolyze urea in the body; the adsorbent would take up the ammonia formed. The artificial cells would, of course, be excreted and would have to be ingested periodically.

Animal experiments have shown that the urease-containing artificial cells can efficiently decompose systemic urea. For practical applications, ammonia absorbents with a much higher capacity are needed.

Another major problem in the artificial kidney is amenable to enzymic correction. Blood clots tend to form in the cannulae which remove the blood from the body and return the dialyzed blood. Streptokinase and urokinase have been used to dissolve the clots. Gonzales and Cocke at Louisiana State University used streptokinase to declot malfunctioning cannulae and found that 17 of 18 cannulae which had been unsalvageable by usual declotting techniques were declotted. Because of the antigenic nature of streptokinase, its reuse was limited and it was suggested that the enzyme should be reserved only for cases in which routine declotting methods have failed.

USE OF AN ENZYME IN THE "ARTIFICIAL PANCREAS"

The pancreas of normal persons releases insulin throughout the day in small bursts in response to increases in blood sugar. Present-day therapeutic methods deliver, to the diabetic, doses of insulin which are several times greater than that delivered by the healthy pancreas. These doses are delivered at arbitrary times, rather than "as needed." Thus, the patient experiences a fluctuation between high and low insulin levels in the blood. It has been suggested that the resulting metabolic stress may be responsible for much of the deterioration in the "well-treated" diabetic.

Drs. S. P. Bessman and R. D. Schultz at the University of Southern California Medical School have described a device, which can be implanted in the body of diabetics, to monitor the blood sugar level and to control a self-contained unit which meters out insulin in proportion to the need as indicated by the blood sugar level. Requirements of the device are: (1) a sensor for blood sugar, (2) an amplifier and computer,

(3) a power supply, (4) an insulin reservoir that could be refilled conveniently from outside the body every 3 months or more, and (5) an insulin pump activated by the sensor-amplifier computer system. Much of the technology for the device is available; the needs are for a reliable, long-lived sugar sensor and for a way of making the implant biocompatible.

One approach to the problem of the sugar sensor makes use of the enzyme, glucose oxidase. Bessman and Schultz converted an oxygen-sensitive electrode to a glucose sensor by covering the electrode with a cloth containing cross-linked glucose oxidase. (The sensor is, in fact, an enzyme electrode.) The enzyme matrix lost less than 10% of its response to physiological concentrations of glucose after subcutaneous implantation in a rabbit for 11 days, and only about 50% of its activity after 6 months.

Glucose and oxygen, dissolved in the blood, react in the enzyme layer to decrease oxygen diffusion to the underlying oxygen-sensitive electrode. An enzyme-free control cell senses the constant oxygen level of the blood. The decrease of current output of the enzyme electrode relative to the control electrode is a measure of the glucose concentration in the blood. The device is highly sensitive to glucose concentrations in the physiological range (50 to 150 milligrams per milliliter). The sensor could be implanted in the bloodstream or under the skin. At present, biocompatibility problems are delaying the use of the device. Platelets (blood cells) and protein tend to accumulate on the sensor, and they gradually choke off the supply of glucose and oxygen to the sensing surface.

Pregastric Esterase in Treatment of Calf Scours

It seems appropriate to include an example of enzymic therapy in veterinary medicine. Drs. J. H. Nelson and M. G. Farnham of Dairyland Food Laboratories have described (and patented) the enzyme treatment of "scours" in animals.

This disease, and its complications, is a major cause of death in calves. Scours are characterized by watery, fetid diarrhea and severe inflammation of the intestinal tract caused by infection or irritating food. Scouring calves become excessively dehydrated as indicated by sunken eyes and an abrupt rise in body temperature. Feces are yellow to white and chalky in appearance. Death may occur within 24 hours, or the disease may continue for a week or more, leaving the animal stunted. The disease is highly infectious and therefore a threat to entire herds.

A wide variety of drugs has been used successfully in the treatment of scours, but inconsistent results have been seen with all of them. The work of the Dairyland scientists has provided a new therapeutic agent which may become the drug of choice.

A controlled study was reported. Ninety-three milk-fed calves that had active scours were treated. Twenty-four received placebos; 29% recovered; the rest became chronic, or died. Forty-two received pregastric esterase in combination with oxytetracycline (an antibiotic); 97% recovered in an average of 2.2 days. Fifteen calves received pregastric esterase alone; 86% recovered in an average of 3.7 days.

Statistical analysis of the data indicates that the enzyme alone is highly effective in curing scours whereas the combination of enzyme and antibiotic is significantly more effective. The enzyme is a valuable drug for treating scours. Antibiotics, in continuous use, can lose effectiveness due to development of mutant bacterial strains that become resistant to the drugs. The enzyme can be used without antibiotics and, when used with them, may be expected to hasten recovery, decreasing the likelihood of development of resistance to the antibiotics.

SOME OTHER MEDICAL USES OF ENZYMES

A recent report mentioned that two enzymes, catechol 2,3–oxygenase and catechol–3,5–oxygenase, are able to detoxify urushiol, the irritant found in poison ivy, poison oak, and poison sumac. The enzymes may some day be used to treat, or prevent, these bothersome conditions.

It occurs to me that critics of this book may say that it suggests enzymes as cures for everything from cancer to baldness. With the charitable intent that such critics be saved from exaggeration, a recent French discovery is mentioned. Guy S. Laporte reported that a lotion consisting of soluble proteins and enzymes was tested in the treatment of certain forms of alopecia (that's baldness, folks!). The lotion was 47.2% effective in arresting or diminishing hair loss.

But it seems appropriate to end this chapter on a more serious note. A recent report in the popular press mentioned the discovery that there is a strong correlation between the "sudden infant death syndrome" and low levels of the enzyme phosphoenolpyruvate carboxykinase (PEPCK) in the liver. University of Wisconsin biochemist, Dr. Henry Lardy, the discoverer of the correlation, reported his findings at the First Annual Symposium on Sudden Infant Death sponsored by the National Institutes of Child Health. The enzyme, PEPCK, is part of the system which converts amino acids to glucose to maintain blood sugar levels between meals. Crib death samples contained, on the average, about one-sixth of the PEPCK activity of samples taken from infants who died from other causes. Samples were taken from 87 crib death victims. Twenty-six of the infants had no signs of other complications; the other 61 had mild respiratory infections. PEPCK levels were normal in 12 "control" babies, who had died in accidents. The enzyme varies from normal to low values in an assortment of other diseases. In most cases, the enzyme levels were assayed by people who did not know the causes of death. Dr. Lardy has emphasized that the low level of the enzyme has not been established as the cause of death. Yet, several aspects of crib death indicate a failure of the body to maintain normal blood sugar levels. Most deaths occur at night; the incidence seems higher in babies who missed a meal or who vomited before bedtime. The discovery may lead to the development of testing methods for identifying babies who might be predisposed to crib death. The "sudden infant death syndrome" may ultimately be found to be an enzyme deficiency disease, like those we shall consider in Chapter 13.

Chapter Eleven
Enzymes Versus Cancer[1]

THE USE OF ENZYMES in the treatment of cancer may be traced to a period preceding the discovery of America. It has been said that the medicine men of the American Indians used papaya leaf and fruit applications to treat malignant diseases, as well as inflammations and infections. Any benefit was doubtless due to the enzyme papain. In the nineteenth century, the enzyme pepsin was used in cancer therapy—first in, the form of stomach juice applied to a surface cancer, and later, by injecting the enzyme into a vein or muscle.

Early in the present century, John Beard, a Scotch embryologist (one who studied the process by which a fertilized egg is transformed into a living creature), treated cancer by use of a pancreatic extract. Therapeutic successes resulted and Beard wrote a book describing *The Enzyme Treatment of Cancer*. Beard theorized that, during the process of differentiation (the process by which cells are transformed into functional organ cells, such as brain cells, kidney cells, heart cells, or muscle cells), still undifferentiated "sex" cells become lost on their way to the gonads. Multitudes of these cells get stuck in various areas of the animal's body. The cells remain dormant and do not multiply during the

---

[1]Much of the material presented in this chapter was taken from the book *Enzyme Therapy* by Drs. Max Wolf and Karl Ransberger, New York: Vantage Press, 1972. Used by permission of Dr. Ransberger.

lifetime of the animal—unless some form of irritation initiates cell division, forming the functionless island of tissue that we call a tumor.

Beard's work stimulated physicians and scientists around the world to explore this area of cancer therapy. But Beard was well ahead of his time. Enzymology was not to be put on a scientific basis for another half a century. Toxic side effects, including fevers and immunological reactions, were seen. Pancreatic extracts produced for therapy were found more harmful than beneficial. Scientists at that time simply did not know about such things as enzyme instability and the presence of fever-causing impurities (pyrogens) in the preparations. Inevitably, enzyme therapy fell into disrepute.

Surgery and radiation have become the most useful tools for fighting cancer. It is vitally important that these techniques be used early, before the disease has metastasized (spread); it is no less important that the treatment encompass the entire extent of the cancer. Anything less may do more harm than good, since both of these techniques are known to actually cause metastases.

A relatively recent development in the treatment of cancer is chemotherapy. The accidental discovery that a war gas, known as nitrogen mustard, produced selective damage to the lymphatic system and the bone marrow led to its testing as an anticancer agent. Temporary remissions[2] were produced in a substantial number of cases. Nitrogen mustard is one of the family of therapeutic agents known as "alkylating agents." Their effects are somewhat like those of X-rays. They inhibit some forms of cancer, but they can also produce cancer and cause genetic mutations.

Therapeutic chemicals called "antimetabolites" are able to block metabolic pathways leading to the formation of deoxyribonucleic acid (DNA). Their effect seems to be slightly greater in cancer cells than in normal cells. Probably the best known of this class is methotrexate, which has been highly successful in treating a form of cancer known as choriocarcinoma. But this drug may produce severe or fatal reactions without warning.

Alkaloids, hormones, and antibiotics have been used in treating some forms of cancer, but there have been no spectacular results. Cur-

---

[2]A remission is defined in medical dictionaries as a diminution or abatement of the symptoms of a disease, or a period in which symptoms are absent or greatly diminished.

rently, it is the usual practice to employ surgery, radiation, and chemotherapy in various combinations to achieve a remission and to sustain it. No one form of therapy is useful against all cancers and every form has its drawbacks. It was to establish this point that we digressed from our consideration of enzyme therapy.

It is useful to resume our historical outline of the development of enzymic therapy of cancer in 1934. In that year, Freund, in Vienna, found that there was a substance in the blood of people who did not have cancer which was able to dissolve cancer cells; the blood of cancer patients did not have this substance. Freund and Kaminer found that the serum and urine of cancer patients was not only lacking in this cancer-dissolving substance, but that cancer cells were even protected against this dissolution by normal serum. When cancer serum was added to twice its volume of normal serum, the cancer-splitting ability of normal serum was lost. Freund was able to isolate this substance, which was water soluble and unstable to heat, from the serum and urine of cancer-free men and horses. He called it "normal substance" and used it with some success in cancer therapy.

Christiani later identified "normal substance" as a cell-splitting enzyme. The inhibitors of this enzyme were cholesterol esters, such as cholesterol succinate and cholesterol butyrate. These are the compounds in cancer patients which counteract the effect of "normal substance."

Dr. Max Wolf identified "normal substance" as an enzyme which decomposes proteins and fatty materials.

The serum of healthy humans and animals is rich in enzymes capable of decomposing proteins, fats, and carbohydrates. Patients having inflammatory diseases and infections generally have decreased enzymic activity; the lowest enzyme content is generally found in the serum of cancer patients. In precancerous conditions, and in early stages of the disease, serum enzyme levels are reduced. It has been deduced from this that low serum enzyme values may represent a predisposition for malignant processes. (These enzymes are quite distinct from the ones that are used to diagnose cancer.) Proteolytic enzymes seem to be of greatest value in revealing predispositions to cancer. Levels of these enzymes decrease in the sickly and aged. This decrease in proteolytic enzyme activity may well be tantamount to a decreased resistance to cancer.

Gaschler resumed the use of enzymes in cancer therapy in 1948, using a product he called "Carzodelan." The preparation was not active when given orally, and it had to be given by other routes. It was not an optimized preparation and therapeutic results were not overly impressive. But it was a start.

The previously mentioned Dr. Max Wolf founded the Biological Research Institute in New York City and there he studied various enzymes which had a selective activity against cancer cells. An interesting tissue culture study was reported by Dr. Wolf.

Cancer cells were grown in close juxtaposition to normal cells. In control samples, balanced salt solutions were used in the culture medium; in experimental samples, enzyme solutions replaced the salt solutions. In the control samples, the cancer cells infiltrated into the normal cells, gradually pressing them back. After a few days, the normal cells showed signs of dying and decomposing, while the cancer cells grew without interference. Events observed in the experimental samples were dramatically different. At first, cancer cells grew without restraint in all directions. Fastest growth was into the direction of normal cells. "Almost suddenly," the growth of cancer cells stopped. Their shapes changed, some shriveled, and finally dissolved. Normal tissues showed hardly any effect of enzyme activity. They pushed back the front rows of tumor cells. Damage to cells in the experimental samples was much more pronounced than that seen in normal tissues in the controls.

After studying various enzymes and combinations of enzymes, Dr. Wolf found that an optimal preparation was obtained by use of a fractionated hydrolysate of beef pancreas, calf thymus, Pisum sativum, Lens esculenta, papaya, and mannitol.[3]

Extensive experiments were performed using rats to determine what routes were appropriate for getting the enzyme mixture into the bloodstream. The enzyme mixture was given to the animals orally, by injection into a muscle or into the abdominal cavity, or by way of the rectum. Serum levels of enzymes were determined 90 minutes after introducing the protein mixtures into the animals' bodies. All of the routes mentioned were effective.

The enzyme mixture, which was found optimal for destroying cancer cells, is available commercially from a German manufacturer. It

---

[3]Wobe-Mugos, the preparation commonly used in Europe for cancer therapy, is described as an enzyme preparation of animal and plant origin. The enzymes are obtained from beef pancreas, calf thymus gland, and the plants Pisum sativum (a variety of peas), Lens esculenta (a variety of lentils) and the papaya tree. Mannitol is not an enzyme. It is a carbohydrate used as a carrier substance for the enzymes.

is sold under the trade name "Wobe-Mugos." We shall have a good deal more to say about this preparation, but it is useful at this point to digress to discussions of some of the properties of cancer cells and the process of metastasis.

SOME PROPERTIES OF CANCER CELLS

If the fundamental differences between normal cells and cancer cells were well known and understood, it would be possible to take a more rational approach to the development of therapeutic agents. Unfortunately, scientists till now have failed to solve this biochemical mystery. Nevertheless, a brief summary of some of the known differences seems useful.

Most of the differences we shall discuss relate to the cell surface. "Adhesiveness" and "stickiness" are two properties that differ. Adhesiveness is a force which holds together attached pairs of cells; stickiness is the tendency of cells to cling to foreign substances such as different cells or glass plates. Normal cells are strongly adhesive, but not very sticky; conversely, cancer cells are very sticky, but not very adhesive.

Two more fundamental differences have been proposed as explanations for these diversities. Cancer cells seem to have a deficiency of calcium ions and they have an overall negative electrical charge. The average charge density of tumor cells is almost twice that of the normal cells from which they are derived. As malignancy increases, the difference in charge also increases. Thus, the fundamental physical law that like charges repel provides an explanation for the lack of adhesiveness. It is quite possible that the deficiency in positively charged calcium ions is responsible for the negative charge. If the normal cell has negative charges balanced by positively charged calcium ions, a deficiency of this ion would leave negative charges unbalanced on the cell. The question then arises: why do cells lose calcium in the process of becoming cancerous. The answer to this is unknown.

Staining experiments by Sagiroglu have revealed that many cancer cells have membrane defects near the nucleus—defects not seen in normal cells. This difference has important implications in enzyme therapy, and we shall have occasion to mention it again.

The classical picture, and the popular notion, of a cancer cell is one of a "rapidly proliferating" cell. Like so many popular notions, it is incorrect. Baserga, in 1965, presented evidence which indicated that there is no constant increase in cell proliferation in malignant diseases. In healthy tissues, cell proliferation is balanced by an equivalent number of cell deaths. It now appears that it is a decreased number of cell deaths, rather than an increased number of cell divisions, that is responsible for the growth of cancers. Thus, instead of being a disease of cellular proliferation, cancer seems to be a disease of cellular accumulation.

The Process of Metastasis

It is often stressed by the American Cancer Society and other cancer fighters that it is important to diagnose malignant diseases as early as possible and to start treatment immediately. An important reason for the stress is provided by the fact that a nonmetastasized cancer is relatively easy to "cure." The primary tumor is not generally the cause of death; more frequently, it is the metastases, resulting from the original tumor, that are lethal.

Cancer researchers have found that the process of metastasis can be considered to occur in four phases:

1. The shedding of individual cancer cells, or groups of cells, from the primary tumor, and their entering the smallest vessels.
2. The transport of these cells through the blood or lymphatic system.
3. The invasion of cancer cells into organs.
4. The proliferation of these cells under suitable conditions with succeeding autonomous growth of a metastasis.

Since cancer cells have a low affinity for other cancer cells (low adhesiveness), they separate easily from the mass of malignant cells that is the primary tumor. Once separated, these highly mobile cells are able to penetrate the blood vessels (usually the veins) to be carried to a site remote from the original tumor, most frequently, to the liver, lungs, or bones.

Malignant infiltration may also occur without the shedding of cells from a primary tumor. The growth of the tumor puts pressure on the peripheral cells of the tumor. In response to this pressure, the cells force themselves into surrounding tissues.

The increased stickiness of cancer cells is extremely important in determining the rate of metastasis, since it favors the lodging of the cells at remote sites to initiate the process.

Kojima has found that the enzyme trypsin is able to reduce the stickiness of cancer cells markedly. A number of researchers have reported that tumors have a high level of the clot-forming protein, fibrin.

The effect of trypsin in reducing stickiness suggests that a surface coating of fibrin is responsible for the stickiness of cancer cells.

The discovery of this trypsin effect was followed by studies of the effects of anticoagulants and of the antithrombin enzyme plasmin. Both the anticoagulants and the enzyme were found to have a cell-poisoning effect on malignant cells. Dr. Max Wolf and co-workers used a combination of animal and plant enzymes (Wobe-Mugos) as a fibrinolytically active agent in experimental animals. The enzyme mixture effectively extended the lives of animals in which tumor cells were implanted.

In contrast to the agents that decrease the stickiness of cancer cells, X-rays and "therapeutic" agents, such as nitrogen mustard, increase this stickiness. Thus, we have at least a partial explanation for the metastases consequent to such treatments.

WOBE-MUGOS

Dr. Max Wolf worked with Ernst Freund (the discoverer of "normal substance") in Vienna in the early 1930s. It was this association, and the work of John Beard, which sparked Dr. Wolf's interest in the possibilities of treating malignant diseases with enzymes. After many years of laboratory studies and animal experiments, Dr. Wolf and his associates started using proteolytic enzymes regularly in cancer therapy. European cancer patients have been benefiting from this therapy since the mid-1940s. Patients have been referred to the enzyme therapists for this treatment or have decided for themselves that they want this type of therapy. Thus, it would be morally indefensible to "treat" such patients with placebos ("sugar pills") in order to provide statistical data. Double-blind studies thus are not available to appease those not impressed by voluminous examples of benefits to desperately sick people. Nevertheless, an attempt was made to obtain statistical data from the results of enzyme therapy performed at more than 46 university clinics and other large hospitals. As explained by Wolf and Ransberger, such attempts were unsuccessful for two fundamental reasons. First, to have a lasting effect, the therapy must be continued over a long period; after discharge from the hospital or clinic, patients tend to stop therapy, especially if not informed of the gravity of such a step. Secondly, enzyme therapy is frequently used in conjunction with other therapeutic measures, and it is thus not certain which therapy is responsible for success or failure.

Prior to 1959, Wolf and his colleagues used the more common

proteases either singly or in combinations. These included trypsin, chymotrypsin, ficin, papain, bromelin, and fungal proteases. Nonproteolytic enzymes also used were deoxyribonuclease, lipase, and extracts from calves' thymus.

The enzyme preparation they used since 1959 is the previously mentioned mixture of enzymes known as Wobe-Mugos. Reports of very good therapeutic results against various forms of cancer have been provided by a number of European clinicians of impressive reputation. These reports all indicate that the therapy has neither disagreeable side effects, nor undesirable late reactions. There was also unanimous agreement that the therapy can be effective only if started early, if the dosage is sufficiently high, and if therapy is continued after the apparent subsidence of active illness.

The composition of the Wobe-Mugos is important. The abundant proteolytic activity raises the fibrin-dissolving power of the serum, thereby eliminating one of the most important conditions for metastases to occur. The lipolytic (fat-dissolving) activity of the preparation is important because cell membranes consist mainly of fatty materials combined with phosphorus and sugars. We have already seen that there is evidence that indicates defects in cancer cell membranes. These defects represent a point of entry, permitting the cell-destroying enzymes to gain access to the cellular material. The fat-dissolving enzymes further attack the membrane facilitating the entrance of the proteases. It is believed that lysosomal membranes are broken by the therapeutic enzymes, thereby releasing additional enzymes from these subcellular particles to join the attack on malignant tissues.

It is quite remarkable, but true, that this enzyme preparation, which can completely dissolve tumor tissue, respects healthy tissue perfectly. The destructive action is aimed only at malignant tissues; there is no attack on normal healthy tissue.

Although Wobe-Mugos is effective when given orally, rectally, or by injection into a muscle or the abdominal cavity, it is most effective when it can be injected directly into a tumor. Whenever possible, this route is preferred. The enzyme mixture is injected in small doses initially, along with lidocaine—a local anesthetic. The dosage is increased gradually. Occasionally, circulatory sensations are noticed, which, in a few cases, may lead to shock. The cause of these sensations is apparently some toxic substance released by the decomposing tumor. If the same dose of Wobe-Mugos is injected beside the tumor or in any other part of the body, such reactions are never seen. The complication, when it occurs, is rapidly reversible and can always be controlled by use of a corticosteroid and a circulatory drug.

It is, of course, pertinent to see what benefits arise from this radically different, nontoxic form of cancer chemotherapy. Not every tumor can be dissolved by injection of the enzyme, but it does seem likely that every cancer patient can benefit in some way from enzyme treatment.

Probably the second worst news that can be given to a cancer patient is that his cancer has metastasized. Numerous clinical publications and reports have established Wobe-Mugos as a most effective agent for the prevention of metastasis. The enzyme mixture shows a number of actions to prevent metastasis. First, it attacks the malignant cells which have been detached from the primary tumor; then it decomposes small clots, in which cancer cells accumulate as a preliminary step to metastasis; finally, it reduces the characteristic stickiness of cancer cells—a property prerequisite for metastasis formation.

We have seen that the primary cancer-fighting tools of today—surgery and radiation—both are attended with the danger of causing metastases. If Wobe-Mugos is used in conjunction with these tools, this danger is dramatically reduced.

The most experienced and most famous cancer hospital in Germany is the Radiation Hospital Janker in Bonn. Dr. Hans Hoefer-Janker, Director of the Clinic, and Dr. Wolfgang Scheef, Chief Physician, reported to the German Ministry of Health on their experience with Wobe-Mugos. The report described how this enzyme preparation increased tolerance to radiation to the point where the usual steroid therapy was most frequently unnecessary, pain-suppressing remedies were rarely needed, and opiates were never needed. One typically long German sentence says much for the therapy: "Only after we could produce a success quota of about 50% by the intratumoral injections, with extreme liquefication necroses of tumors or their metastases, and in several cases, we were able to bring tumors of the size of 'babyheads,' which were beyond any further therapy by cytostatics or radiation, to a complete remission, only then we became fully convinced that this enzyme preparation does not represent just an adjuvant in the tumor therapy, but that it has even to be enclosed in the small group of really causal therapeutics."

Other clinicians have found that tumors are more sensitive to radiation when enzyme therapy is used; the shrinking of the tumors is accelerated. Enzyme therapy supports radiation treatment, and, unlike other chemical agents, it produces no toxic effects. Overdoses are not possible.

The enzyme mixture is useful in conjunction with surgery as well. Postoperative hemorrhage is avoided and, just as in radiation therapy, strong pain killers are unnecessary. The antitumor activity of the preparation causes the malignant mass to shrink and inoperable patients are sometimes brought to a condition where an operation becomes feasible.

Wobe-Mugos extends the life expectancy of the cancer patient, and, once remission is achieved, recurrences are most unusual, as long as the patient continues enzyme therapy. Patients who have become resistant to other forms of therapy are helped by Wobe-Mugos.

There are, of course, patients who cannot be saved by any therapy. Even these are helped by the enzyme preparation. Numerous reports indicate that suffering is greatly reduced in terminal patients, so that the need of analgesics and opiates is lessened and patients do not die in agony. One report mentioned 17 patients who had died. Nine of these were able to leave their beds up to 3 days before death; two died while taking walks.

WOBE-MUGOS AND VITAMIN A[4]

Massive doses of Vitamin A have been found beneficial in cancer therapy. In order to provide high levels of this fat-soluble vitamin in usable form, the Mucos Co. developed a vitamin A emulsion that is easily taken and absorbed. At a symposium on the use of vitamin A in tumor therapy, sponsored by the Cancer Research Institute of Vienna University, Dr. Hoefer-Janker told of an American radiologist (a physician specializing in radiation therapy), suffering from a metastasized cancer, who refused the surgical treatment suggested in the United States. At the Janker Clinic, he refused radiation, insisting on vitamin A therapy. His tumor was completely removed by vitamin A therapy alone.

Vitamin A seems to exert some of its anticancer effect by releasing enzymes from lysosomes (subcellular particles). This mode of action suggested a combination of vitamin A and enzyme therapy. The vitamin enhances therapy by enzymes, radiation, and alkylating agents. It ap-

---

[4]There are potentially serious side effects associated with vitamin A therapy. It can only be considered in a hospital. Massive doses are used—doses which are best administered in the emulsified form available in Germany. Special preparations of the vitamin are mandatory.

pears to act synergistically with Wobe-Mugos, and it has been suggested that proteolytic enzymes should be given at the highest possible doses during the entire duration of vitamin A therapy. The use of Wobe-Mugos was discussed at the symposium on vitamin A therapy in Vienna. Dr. Heinrich Wrba, Director of the Cancer Research Institute, closed the symposium speaking of Wobe-Mugos. He found it hardly understandable that such a nontoxic and, at the same time, effective medicament should be withheld from patients simply because of neglect or lack of interest.

A Recapitulation

We are now prepared to summarize the benefits of Wobe-Mugos. It acts directly on tumors, frequently dissolving them. It substantially reduces the occurrence of metastases and recurrence of cancer. It makes malignant tissues more sensitive to radiation and, at the same time, it reduces pain and other side effects of radiation. When used in conjunction with surgery, it reduces postoperative pain and hemorrhage. In terminal patients, it eases the pain of the last days, eliminating the need for strong drugs, which can make the patient somewhat less than human. It is useful in conjunction with many other forms of therapy. It is effective against a broad spectrum of malignant diseases including cancers of the reproductive organs, the breast, the skin, the digestive system, connective tissues, and others including Hodgkins disease and leukemia. Many cancer researchers are vying for the honor of being the first to prove that cancer is caused by a virus. The enzyme preparations available from the Mucos Co. are effective against viral diseases including those caused by herpes zoster and herpes simplex. The latter is widely suspected of causing cancer. In stark contrast to every other known form of cancer therapy, this enzyme therapy has no known adverse effects!!! *It does not produce problems; it only helps.*

Conclusion

A recent (1976) issue of *Postgraduate Medicine* carried a statement that more than 95% of all cancer cured in the United States and throughout the world, is cured by surgery, radiation, or a combination of the two. This leaves a scant 5% for chemotherapy. Billions of dollars have been spent on cancer research in the United States alone. Literally hundreds of thousands of compounds have been synthesized to be tested for therapeutic effectiveness against neoplastic diseases. A few dozen of these, at most, are used clinically, and all have serious toxic side effects.

Were it not for the gravity of the disease they are intended to fight, their use could not be considered. The low success rate of the "best" compounds that have been used suggests that a different approach to the cancer problem is desirable. We are indeed fortunate in having available an alternative form of therapy which surpasses the old therapy, in that it is more effective and is completely safe.

The Austrian Medical Association held a meeting in Vienna in December of 1971. Professor Heinrich Wrba, head of Austrian Cancer Research Institute at the University of Vienna, at the close of the meeting said in part, "our present knowledge allows us to include this (enzyme) therapy into the small list of highly effective causal anticancer compounds. It will certainly play an important role in cancer treatment of the coming years. Everyone who is really sincere in his efforts to improve the thus far discouraging results of cancer therapy has to intensively engage himself and learn about this most interesting and promising new field of therapy."

We live in an era that has produced few men who can legitimately be considered heroes. Dr. Max Wolf was one such man. Dr. Wolf died on June 3, 1976, without having had the satisfaction of seeing American cancer patients benefitting from the enzyme therapy which he worked to perfect.

This chapter is dedicated to the memory of Dr. Wolf in the hope that it will hasten the day when enzyme therapy is available in the U.S. for treating cancer patients.[5]

---

[5]After this book was submitted for publication, the F.D.A. gave permission for the sale of Wobe-Mugos, Wobe Enzyme and Vitamin A emulsions in the U.S. The materials may be sold without any claim. They are available from the Mucos Pharmaceuticals Corporation of Los Angeles, California.

Chapter Twelve

L-Asparaginase— The Antileukemia Enzyme

LEUKEMIA IS A form of cancer characterized by excessive proliferation of certain types of blood cells. Acute leukemia in man is usually in an advanced stage by the time it is diagnosed. It is estimated that there are 10^{11} to 10^{13} leukemic cells present, depending on the patient's age and the extent of the disease.[1] Under most circumstances, reducing the number of these cells by a factor of 100 to 1,000 brings about a "complete remission." This means only that the cancer cells remaining cannot be detected by the means ordinarily used. There may be a billion or more leukemic cells remaining in a patient in complete remission. Once remission has been achieved, elimination of all these surviving leukemic cells becomes the therapeutic objective. Still, the first step is obtaining the remission.

Spontaneous remissions (without therapy) are sometimes seen —often in association with a fever. This and the fact that some cancer patients show some resistance to the progress of the disease has led scientists to believe that the body's immunological system may represent a defense against cancer. The patient's defenses are set to eliminate any tissue recognized as "nonself" or foreign. Lymphocytes (a kind of white blood cell) recognize the cancer cells as foreign and produce special protein molecules which react specifically with the materials which

---

[1] $10^{13} = 10,000,000,000,000 =$ ten trillion.

caused their formation. These special proteins are called antibodies and the foreign materials which bring about their production are antigens. The antigen-antibody adduct is eliminated from the body.

Some scientists believe that malignant tumors can develop only when this immunological defense is weakened. Generally, malignant diseases develop at extremes of life when the defense is weak. Those suffering from immunologic deficiencies have an increased risk of malignant diseases. Transplant patients given drugs to ward off rejection (immunosuppressant drugs) have a greatly increased incidence of cancer.

Since 1895, researchers have sought ways to strengthen the body's defense against malignant diseases. Over the last hundred years, it has been found possible to boost the immunological system's defense against diseases by injection of an antiserum.

Early in the 1950s, Dr. John Kidd of Cornell University Medical School conducted experiments aimed at developing an antiserum to leukemia. Mice used in his work were inbred for many generations so their genetic makeup was uniform. Thus, tissues could be taken from one of the mice and transplanted to another without fear of rejection. Leukemia and other cancers were transplanted into these mice. Potential anticancer agents were then tested in several animals carrying the same transplanted malignancy.

Leukemia cells from these mice were transplanted into rabbits in an attempt to get the desired antiserum to leukemia. The rabbit's body should recognize the foreign materials and synthesize the antibodies necessary to reject them. The rabbit serum would then be the desired antiserum. It would be taken from the rabbit and injected into the leukemic mice to see if it would bring about the desired rejection of the leukemia.

A group of proteins known collectively as "complement" enhances the immunological reaction. Guinea pig serum is a good source of complement and Kidd therefore used a combination of the rabbit antiserum and guinea pig serum in his attempts to cure the leukemia transplanted into the mice. As a control, some of the mice were injected with guinea pig serum only—no rabbit "antiserum" was given. It was expected that these control mice would show continued growth of the transplanted leukemia and would die from it, while those given the antiserum would show improvement.

The combination of the rabbit antiserum and the normal guinea pig serum was very effective in controlling the implanted leukemia. A startling result was seen in the control mice—those injected with guinea pig serum alone. Their leukemias regressed and, in some instances, were cleared completely.

Kidd tried sera taken from rabbits, horses, and humans to see if these would bring about the same results produced by the guinea pig serum. No other serum he tried was effective. He established that guinea pig serum affected only malignant cells leaving normal cells unaltered and producing no toxic side effects. No pathological changes in the animal's tissues could be seen by the naked eye or by microscopic examination. The guinea pig serum was very effective against some transplanted leukemias, but was without effect on others. No newly transplanted leukemias were sensitive to the serum.

This latter observation was unfortunate since it probably delayed by several years the arousal of widespread interest in the phenomenon. Because it seemed that only long-transplanted leukemias were sensitive and because of the way it was discovered, many believed that guinea pig serum acted, through its high level of complement, to augment an antigen-antibody reaction between the host mouse and the transplanted leukemia. Nevertheless, as Kidd pointed out, his discovery provided a unique example of a naturally occurring substance that brings about regression of cancer cells in living animals without doing obvious harm to the animals.

Kidd carried out further studies to try to determine what component of guinea pig serum was responsible for the antileukemia activity. He separated the complement proteins into various fractions to see if he could find one specific group of proteins which caused remission of leukemia. None of the fractions had this property.

Amino Acids Essential in Cancer

Tissue culture is a technique used extensively in biological, biochemical, and medical research. Living tissue is kept alive by incubating it in a broth containing all the nutritional materials needed for the survival of the cells. By using this technique, it is easy to determine what nutrients are essential for the cell's survival.

Not long after Kidd's publication of his discoveries, researchers at the Noble Foundation in Ardmore, Oklahoma, reported that some rat tumor cells, maintained in tissue culture, died unless they were supplied with two amino acids which are not essential to normal cells. These amino acids were L–asparagine and L–glutamine. This was the first report of a nutritional element that was essential to a tumor cell but not to a normal cell. No connection was perceived between this observation and Kidd's discoveries.

Identification of the Antileukemia Factor

Dr. John Broome, a British scientist working in Kidd's laboratory at Cornell, was convinced that the antigen-antibody explanation for the tumor-inhibiting activity of guinea pig serum was untenable. His search for a nonimmunological difference between guinea pig serum and other sera was started in the library. A paper published in 1922 provided a valuable clue. A scientist named Clementi, working at the University of Rome, had found that an enzyme, L–asparaginase, was present in the liver and kidneys of a wide variety of species, but was found in the blood of only one—the guinea pig.

L–asparaginase decomposes the amino acid L–asparagine, forming aspartic acid and ammonia. (Recall that L–asparagine was found to be essential to some rat tumors although it is not essential to normal tissues.)

The Clementi report prompted Broome to compare the physical and chemical properties of guinea pig serum L–asparaginase with those of the antileukemia component of the serum. Asparaginase activity could be determined by measuring the amount of ammonia released when the enzyme was incubated with excess L–asparagine. The antileukemia activity was measured by injecting the material into leukemic mice and looking for a decrease in cancerous cells.

Both the enzyme activity and the ability to destroy leukemias were maintained when heated to 56° C for half an hour. Both were destroyed when heated to 66° C. Both were rapidly inactivated when the pH was lowered to 4.4, but were more stable to alkaline solutions up to pH 11.2. The blood of newborn guinea pigs had been found to have no tumor-inhibiting properties; it was found to contain very little L–asparaginase.

Broome then undertook the purification of guinea pig serum asparaginase and he tested the various fractions to see if they related to tumor inhibition. Both the enzyme and the tumor-inhibiting component were precipitated in a 15% solution of sodium sulfate. When the serum components were separated by using an electrical technique (electrophoresis), both activities were found in a single peak. It was concluded that L–asparaginase was, in fact, the component responsible for the antileukemia activity of guinea pig serum.

Broome's purificaton gave a product having 118 times as much activity as the starting material. Yellin and Wriston at the University of Delaware used a somewhat more complex procedure and achieved a 900-fold increase in asparaginase activity. They found the enzyme to be highly inhibitory to those tumors which were sensitive to guinea pig serum.

Interest in L–asparaginase was aroused by Broome's report and new sources of the enzyme were sought. A number of South American rodents (most notably the agouti) were found to have high levels of serum asparaginase and these sera were more powerfully inhibitory to tumors. The enzyme was also found in two species of "old-world" monkeys, but not in "new-world" species.

It was soon realized, however, that guinea pig serum was the only common animal source of the enzyme and that millions of guinea pigs would be needed just to supply the enzyme if the leukemia-inhibiting property was to be exploited. Microbial sources of asparaginase were sought since such sources could potentially provide unlimited supplies of the enzyme. Asparaginase from yeasts was totally devoid of cancer-inhibiting properties. This was attributed to the fact that yeast asparaginase was almost completely cleared from the blood within 30 minutes after injection. Half of the enzyme activity from guinea pig serum could be found in the blood after as long as 19 hours.

A number of other microbial asparaginases were found to be without cancer-inhibiting properties, but, in 1964, Mashburn and Wriston reported that a microorganism called Escherichia coli (usually abbreviated E. coli) produces an asparaginase which has greater antileukemia activity than the guinea pig enzyme. It was later found that E. coli produces two distinct types of asparaginase, one of which is highly active against certain tumors, while the other has no tumor-inhibiting activity. The tumor-inhibiting enzyme was found to have a high avidity for asparagine, while the other enzyme had low avidity for the substrate.

The discovery of a highly active E. coli asparaginase—an enzyme which could be isolated and purified on a large scale—made it possible to test a wide variety of cancers for sensitivity to the enzyme. Mice, rats, and dogs were the usual experimental animals. Only about one cancer in four was sensitive to asparaginase.

Dogs were particularly interesting experimental animals because they belong to an order of animals (the carnivores) far enough removed from the order of rodents to provide a significant test of how general the occurrence of asparaginase-sensitive cancer is. Dogs also have a high incidence of spontaneous (nontransplanted) lymphosarcoma, which resembles some human cancers.

Dr. L. J. Old and associates tried asparaginase therapy on three dogs having far-advanced lymphosarcoma. The dogs' lymph nodes and tonsils were greatly enlarged and they could hardly move. Therapeutic results were very encouraging. Within a week of the start of therapy, two of the dogs were restored to apparently normal health and the third showed considerable improvement. In the words of Dr. Old, "It is remarkable that a cancerous tissue can be destroyed on such a scale, with concomitant release of breakdown products, without evidence of toxicity." The technology required for production of large quantities of highly purified asparaginase was not sufficiently developed to permit studies of the effects of massive doses of the enzyme or of extended periods of treatment. The lymphosarcomas reappeared in the dogs between 7 and 50 days after termination of therapy. One of the dogs was treated again with asparaginase and a second remission was obtained.

The first use of asparaginase in primates was reported by Dolowy and associates in 1966. After careful examination of the effects of partially purified guinea pig asparaginase on mice, the enzyme was injected into three normal (noncancerous) female rhesus monkeys. The dose given was that estimated to be the maximal initial dose to be used in humans. During the course of the intravenous injections of the enzyme, there was no increase in blood pressure, no change in heart rate or respiratory rate, and no jaundice. In short, these investigators did not see any of the toxic effects which were considered probable. The animals remained alert and were neither overly active nor lethargic. In fact, the only significant changes noted were an increase in blood urea nitrogen noticed 24 hours after injection and a transient decrease in circulating lymphocytes. Both returned to normal after 9 days.

As we have seen, when asparaginase acts on the substrate asparagine, ammonia and aspartic acid are produced. The Dolowy team recognized the possibility that these decomposition products could be responsible for the antileukemia activity. These and related compounds were injected into tumor-bearing mice. None of these compounds caused any regression in tumor size or a decrease in the median time to the death of the animals.

Treatment of Human Leukemia with Asparaginase

In accordance with pertinent laws and hospital rules, informed consent to asparaginase therapy was obtained from the parents of an 8-

year-old boy with a 3-year history of acute lymphoblastic leukemia. For 30 months after the initial diagnosis, the boy had been treated with conventional antineoplastic (that is, cancer-fighting) drugs and with radiation. He responded with complete, then partial, remissions, but after 33 months none of these treatments was effective. The boy was in relapse with progressive disease. It was at this point that the decision was made to try asparaginase therapy and consent was given.

A number of bad effects were found associated with the injection of guinea pig enzyme into this first patient. There was a rise in temperature, pulse rate, and respiratory rate and a drop in blood pressure. Ten hemorrhagic bowel movements occurred during infusion of the enzyme. The boy was given a transfusion of fresh whole blood. Evidence of marked hemolysis (splitting of blood cells) required cessation of asparaginase therapy.

Favorable results of the therapy included a decrease in total white blood cells, a decrease in lymphoblasts (blood cancer cells), a decrease in liver size, and a decrease in the size of a cancerous testicle.

The patient died with a pulmonary hemorrhage 10 days after the treatment. Leukemic cells were found in several organs.

The first remission of leukemia attributed to asparaginase therapy was reported by Hill and co-workers in 1967. Three patients suffering from acute lymphoblastic leukemia which was not responsive to cytotoxic agents (chemicals poisonous to living cells) were treated with asparaginase. In response to this therapy, two of the patients had a lessening of white blood cells and the third had a complete remission.

Hill found that the acute toxicity and side effects that were common to other antileukemia agents in use at that time were insignificant in asparaginase therapy. This apparent lack of toxicity led Hill to use larger doses of the enzyme. Six of seven acute lymphatic leukemia patients and four of eight acute granulocytic leukemia patients underwent a complete remission when given large doses of the enzyme. Only two of 12 patients with acute lymphatic leukemia and none of 10 with acute granulocytic leukemia obtained remissions when lower doses were used. Other clinicians have reported that the response to asparaginase is not strictly related to dose and that remissions have been obtained using very small doses. While asparaginase has been found to be highly effective in treating many cases of acute lymphatic leukemia, it has been less promising in treating other forms of leukemia and solid tumors.

An acute disease is one that comes on suddenly; this is opposed to the chronic type, which comes on slowly. In lymphatic leukemia, lymphocytes (white blood cells produced in the lymph glands) are abnormal and reproduce wildly. Another form of leukemia is due to abnormal production of granulocytes, which are white blood cells manufactured in the bone marrow. This form of leukemia is usually called granulocytic, but it is also known as myelocytic or myelogenous leukemia.

Toxicity of Asparaginase

While asparaginase does not show the kinds of toxicity associated with the cytotoxic agents, it does have some adverse side effects. Among these effects are nausea, loss of appetite or vomiting, fever (sometimes severe), anemia, blood-clotting abnormalities, liver dysfunction, inflammation of the pancreas, and central nervous system impairment.

Despite what looks to the layman like a frightening list of side effects, asparaginase is tolerated with no or minimal side effects by most patients. The toxicity that is associated with asparaginase is believed to be due to reduction in protein synthesis.

Asparagine is a component of most, if not all, proteins, and, when the blood concentration of this amino acid is reduced to undetectable levels, there cannot be much available for protein synthesis. Asparaginase rapidly reduces the asparagine concentration below these levels. After the decline in protein synthesis, there is a resultant decrease in the synthesis of the genetic materials—deoxyribonucleic acid (DNA) and ribonucleic acid (RNA).

Asparaginase is now used in large doses, always in combination with other antineoplastic agents. It is still more useful in the treatment of acute lymphatic leukemia than in other forms of the disease and is rarely useful against solid tumors.

Mechanism of the Antileukemia
Effect of Asparaginase

It is reasonable to ask why asparaginase should have the ability to fight leukemia. The answer to this has not really been established, but it is generally believed that asparagine is essential to cancer cells, although it is not essential to normal human cells. This belief is supported by the study of the tissue culture involving cancerous rat cells which required asparagine and glutamine for survival. Furthermore, the level of plasma asparagine in leukemia patients before the start of therapy has been found to be lower than that of normal healthy people. This suggests that the leukemia cells have a greater need for asparagine than normal cells and that they get this amino acid from the plasma. (Cancerous cells seem to have a greater avidity for needed nutrients than do normal cells for it is often seen that cancer grows readily in people who are starving.) Capizzi and co-workers at Yale University and the Hospital of St. Raphael at New Haven, Connecticut, found that the level of plasma asparagine correlated well with the disease state in several patients; the level decreased as the disease worsened. This lends further support to the theory that the destruction of asparagine by asparaginase explains the antileukemia activity of the enzyme. Based on the belief that this is the correct mechanism, asparaginase therapy is sometimes called "asparagine depletion therapy."

Other Features of Asparaginase Therapy

Another question that arises when considering asparaginase therapy is why should one leukemia patient be so responsive to the therapy that he achieves complete remission while other patients are not helped by it. The answer seems to involve another enzyme called asparagine synthetase. Drs. Haskell and Canellos at the National Cancer Institute found that before asparaginase therapy, asparagine synthetase was virtually undetectable in human leukemia cells regardless of the eventual outcome of therapy. After therapy, however, the cells from patients who were not helped by asparaginase showed a sevenfold increase in asparagine synthetase levels as compared with those who were helped. It thus appears that some patients experience an induction of asparagine synthetase in response to asparagine depletion. This counteracts the effects of asparaginase. Those patients who are aided by asparaginase do not bring about asparagine synthetase induction.

Still another question arises. Guinea pig serum has been found to produce an antileukemia asparaginase; yeast enzyme is without effect in treating leukemia; E. coli produces two asparaginases—one active against leukemia and the other inactive. Since asparaginase, by definition, is an enzyme which destroys asparagine, why should one asparaginase be active against leukemia while another asparaginase is not?

The answer is that the various enzymes having asparaginase activity have different structures and different amino acid components. Some are huge molecules having molecular weights in the hundreds of thousands; others are considerably smaller and have a molecular weight of about 30,000. There are at least three features of the asparaginase enzymes which result from the differences in structure and which are important in determining whether or not the enzyme will have antileukemia activity. First, since the enzymes will be competing with the leukemia cell for the available asparagine, the enzyme must have a greater affinity for the substrate than the cell has. An enzyme having low affinity for the substrate will be ineffective. Secondly, it must have a low antigenicity to avoid rapid inactivation by antibodies formed against it by the patient. Finally, the enzyme must not be so large as to facilitate its removal from the blood by the body's filtering system.

In regard to the second feature—low antigenicity—it appears that, if asparaginase therapy is continued long enough, antibodies will eventually make asparaginase from any given source ineffective. Fortunately, antibodies formed in response to one active asparaginase do not affect asparaginase derived from a different source. This arises from the highly specific nature of the antigen-antibody reaction and the different structures of the asparaginases derived from various sources. By using asparaginases from a number of different sources, asparagine depletion therapy can be continued long after antibodies are formed against the first type of enzyme used. As will be discussed later, the use of immobilized enzymes can eliminate the possibility of antibody formation.

Immunosuppressive Activity of Asparaginase

While discussing asparaginase in its relation to antibodies, it should be mentioned that this enzyme and a number of related enzymes have been found to be immunosuppressive. That is to say that they reduce the body's ability to recognize foreign proteins and remove them from the bloodstream. Because of this immunosuppressive property, asparaginase has been proposed for use in prolonging the viability of organ transplants. It is the antigen-antibody reaction that causes the rejection of transplants. Asparaginase has already been used successfully to prolong transplants in animals.

The immunosuppressive activity of asparaginase, like its antineoplastic activity, is believed to be due to asparagine depletion and the resultant decrease in protein synthesis. The enzyme causes a delay in the appearance and proliferation of antibody-forming cells and a consequent delay in the production of serum antibody.

Immobilized Asparaginase

Among the problems associated with asparaginase therapy are toxicity, antigenicity, and the use of large doses of relatively expensive enzyme. The use of immobilized asparaginase has been suggested as a means for avoiding these problems. The enzyme has been bonded to many types of soluble, insoluble, and colloidal (gelatinlike) supports. Insoluble carriers have included cellulose, glass, synthetic polymers, and stainless steel. Chang prepared artificial cells consisting of asparaginase in nylon microcapsules. These were more effective than asparaginase solution in suppressing lymphomas implanted in mice.

Nylon tubing has provided an interesting type of enzyme support. The tubing permits unobstructed blood flow while holding the enzyme in place on the inner surface of the tubing and permitting direct contact of the blood with the enzyme. The degree of contact can be reduced by attaching a layer of gel to the inner wall of the tubing. The enzyme is then bonded to the gel, which provides a filtering effect. Small molecules, such as asparagine, can easily penetrate the gel to be broken down by the enzyme, but large protein molecules cannot get in. This prevents the possibility of antibody formation.

Soluble-supported asparaginase derivatives have been prepared by bonding the enzyme to soluble dextran and dextranlike molecules. Certain acidic derivatives have been found to have the useful property of being soluble under certain easily determinable conditions and insoluble under other conditions. This is discussed more fully in Chapter 3.

Asparaginase in an Artificial Kidney

Some useful information on asparagine depletion therapy has been obtained by studying patients being treated with a "kidney machine." The blood asparagine content of these noncancerous patients was measured before "therapy" was started. After 6 hours of hemodialysis, the amount of asparagine removed from the blood was determined. Ten times as much was removed as was found to be present before the treatment. It was concluded that there must have been extensive input of asparagine either from tissue stores or from synthesis of the material by the body during the treatment period.

Hemodialysis has also been carried out using asparaginase in the dialysis fluid. This has produced improvement in the condition of leukemia patients and some degree of regression of solid tumors.

Clinical Use of Immobilized Asparaginase

Dr. Holger Hydén at the University of Göteborg in Sweden seems to have been the first to report the actual clinical use of an arteriovenous shunt containing immobilized asparaginase. Such an approach had been suggested in the earliest papers describing the preparation of immobilized asparaginase for use in leukemia therapy. Hydén used enzyme attached to glass plates or plates made of a synthetic polymer. The shunt was used in animal experiments to determine its effects on the circulation and coagulation of the blood. After such work revealed the safety of the apparatus, it was used to treat a human patient suffering from a malignant lymphoma. The shunt was used over 18 times in 38 days for periods ranging from 0.5 to 6 hours. There was never any discomfort associated with the treatment, no signs of allergy, and no problems with the kidneys, liver, or the circulatory system. Therapeutic effects have not yet been reported, but the procedure has been shown to be safe.

Sampson and co-workers at Roswell Park Memorial Institute in Buffalo studied the effects of insoluble extracorporeal asparaginase on the blood (apparently noncancerous) of patients in end-stage renal failure. Heparin, an anticoagulant, was added to the blood as it left the body and protamine, which neutralizes the heparin, was added as the blood returned to the vein. Patients tolerated the procedure well and no deleterious effects were seen. There was a prompt reduction in serum asparagine and the level remained low as long as perfusion of the blood through the shunt was continued. In noncancerous patients, no therapeutic effects could be expected from the treatment, but it was shown that the toxicity normally associated with asparaginase could be avoided by the use of the extracorporeal method of treatment, that the procedure was safe, and that it did result in asparagine depletion.

Epilogue

Asparaginase therapy has been somewhat disappointing in the sense that it has not provided the long-sought "cancer cure." Nevertheless, asparaginase is one of a number of useful drugs for controlling leukemia, particularly the acute lymphatic form. Asparaginase is used in combination with other drugs, most notably prednisone and vincristine. A large fraction of leukemic lymphoblasts can be killed without serious and extended injury to normal marrow cells by using the three drugs properly. The combination of these three drugs with radiation of the central nervous system has produced long-term survival of victims of acute leukemia. A generation ago, such patients would have had little hope of survival. (The radiation is essential to destroy any leukemia cells "hiding" in the central nervous system where they are protected from drugs acting in the bloodstream. This protection comes from the "blood-brain barrier.")

The story of the accidental discovery of the therapeutic effect of guinea pig serum reminds us that many outstanding discoveries are made by trained people who are looking for something else. The rest of the asparaginase story shows the deliberate step-by-step research that is necessary to convert such a discovery to a therapeutic weapon. The animal studies (called "preclinical pharmacology" by the physician) provide guidance in estimating the correct doses to be used in therapy and are useful in predicting toxic effects. Research continues into the stage of human therapy for it is useful to determine how the drug is distributed in the body and how it is metabolized when given by different routes and in different doses.

It is useful to mention again two features of a research project without which the therapeutic effects of asparaginase might not have been discovered. The control experiment is one of these features. If Kidd had not run his control experiment, the therapeutic effect would have been attributed to the rabbit "antiserum" rather than to the guinea pig serum.

The second feature is the use of the library. The essential clue to the identity of the antileukemia factor in guinea pig serum was found in the library. Without this crucial piece of information, the task of identifying the active component would have been considerably more difficult.

Asparaginase is still the focal point for a good deal of research. With the advent of immobilized enzymes, future research into the clinical use of insolubilized asparaginase is assured. It should be mentioned in closing that researchers are still trying to develop an antiserum to leukemia. Russian scientists have reported the isolation of a leukemia-producing virus. If this report proves correct, the long-sought antiserum will probably follow very quickly.

It may ultimately be found that the combination of asparaginase and glutaminase is more effective than asparaginase alone. This combination would destroy both of the amino acids found essential in some cancer cells and it would remove the substrate needed for asparagine synthetase.

Chapter Thirteen

Enzyme Deficiency Diseases

A PRINCIPAL REASON for writing this book is to show the fundamental importance of enzymes. To this end, many uses of these biological catalysts have been described. In this chapter, we show some of the drastic consequences which may arise when a single enzyme is missing from the human body. A number of techniques used in attempts to relieve the enzyme deficiencies are also described.

All of the enzyme deficiency diseases are genetic diseases. That is, they are all inherited. Except for one, all are inherited as autosomal recessives. This means that children will not have the disease unless both parents are carriers of it. The diseases are equally likely to strike male or female children. On a statistical basis, if both parents are carriers, one child in four is expected to have the disease, two are expected to be carriers like the parents, and the fourth is not expected to be involved in the disease at all.

The only known exception to this mode of inheritance among the enzyme deficiency diseases occurs in Fabry's disease, which is carried only by the mother. It is thus inherited as a "sex-linked recessive." Half of the male children of a Fabry carrier are expected to have the disease and half of the daughters are expected to be carriers. In contrast to the lack of symptoms seen in carriers of other enzyme deficiency diseases, women who carry Fabry's disease may show some of the symptoms although they usually are much milder than those seen in males suffering from the disease.

There are more than 100 inherited diseases which have been investigated thoroughly enough to reveal the nature of the biochemical defects causing them. Not all inherited diseases involve enzyme deficiencies, but this chapter is limited to those that do. Even with this limitation, we can consider only a few of the diseases. Those included were selected because of historical importance or because of some particular feature of the disease or mode of therapy.

When speaking of enzyme deficiencies, we include all cases in which the activity of the enzyme is low enough to result in pathological effects. It is not necessary that the enzyme be completely absent. Scientists believe that defective proteins having greatly reduced enzymic activity may be synthesized by victims of the diseases. In one case, a disease known as acatalasemia, the enzymic activity of catalase (an enzyme which decomposes hydrogen peroxide) is low because an unstable enzyme is synthesized by victims of the disease. Although there may be hundreds or thousands of amino acid residues in a given protein, if only one of these is not the normal one, the enzyme may have reduced activity or no activity at all. Normal hemoglobin (a nonenzyme protein) contains some 574 amino acid residues. Sickle cell anemia is caused by substituting one valine for a glutamic acid residue in an apparently critical part of the hemoglobin molecule.

There is immunological support for the belief that defective proteins may be synthesized by victims of the enzyme deficiency diseases. The antigen-antibody reaction, like the enzyme-substrate reaction, is highly specific. The antibody is able to select its specific antigen in the presence of a great number of other proteins. It will react with that molecule leaving similar molecules unaffected. Scientists have been able to isolate antibodies to some of the enzymes, the lack of which causes deficiency diseases. If the enzyme is missing in an individual, the antibody has nothing with which to react. It has been found that the injection of the antibody into the victims of deficiency diseases causes precipitation of inert proteins. These inert proteins are believed to be the defective enzymes.

Considerable insight into these diseases was provided by Sir Archibald Garrod around the turn of the present century. He described four families having a high incidence of alcaptonuria—a rare disease in which victims excrete urine which turns black if left standing at alkaline pH. Garrod analyzed the urine and found that the black color was due to homogentisic acid—a material not found in normal urine. He believed the disease was caused by a deficiency of the enzyme homogentisic acid oxidase in the liver. This was confirmed 50 years later.

Garrod found a high number of consanguineous marriages (that is, between blood relatives) among patients suffering from alcaptonuria. This feature was also seen in several other diseases. These included albinism, which is characterized by a lack of pigment formation, pentosuria, in which there is a high level of five-carbon sugar in the urine, and cystinuria, in which patients have high urinary levels of the amino acid cystine. He concluded that victims of these diseases suffered from a deficiency of an enzyme required to catalyze a specific step and that this deficiency was inherited. Garrod called such diseases "inborn errors of metabolism." His conclusions, described in a book bearing that name, gave birth to the science of biochemical genetics and to the idea that the synthesis of enzymes was genetically controlled.

It is known today that inborn errors of metabolism are involved in more than 100 human diseases. Some are so mild that no treatment is prescribed. Others are so severe that they cause complete physical and/or mental incapacitation or even a particularly horrible death in infancy.

In most of the enzyme deficiency diseases, the patient suffers from the effects produced by accumulation of the material which would have been broken down by the "missing" enzyme. The materials tend to accumulate in organs such as the kidneys, liver, or spleen or in muscle tissue including the heart or in the central nervous system. Often the site of storage is an organ which is normally very rich in the enzyme which is missing.

The last 25 years have produced a vastly improved ability in the art and science of enzymology. A good deal of research is now aimed at treating these enzyme-related diseases. Dr. Roscoe Brady of the National Institutes of Health has described several stages of research into the enzyme deficiency diseases involving lipid storage. These stages include: (1) recognition of the disease syndromes, the group of symptoms which characterize the disease; (2) identification of the materials that accumulate; (3) elucidation of the nature of the enzyme abnormality; (4) isolation, purification, and examination of the enzyme involved; (5) enzyme replacement therapy; (6) genetic engineering. He feels that stage 5 is just starting and that stage 6 is for the future. Genetic engineering deals with changing parts of the chromosomes. It is anticipated that years of diligent research will be needed to bring such an idea to the point at which clinical trials can be considered.

ACATALASEMIA

Some of the research that is being done in the area of enzyme replacement can be illustrated by considering acatalasemia. This disease was first discovered by Dr. S. Takahara, a Japanese surgeon. During an operation, he applied hydrogen peroxide to a wound and observed that there was no foaming and that a dark brown color was produced. In a normal individual, catalase in the blood decomposes hydrogen peroxide forming water and free oxygen: $2H_2O_2 \qquad 2H_2O$ and O_2.

The liberation of oxygen causes foaming. Acatalasemia is a disease in which there is no catalase activity in the blood. Peroxide is not broken down, so there is no foaming. A dark brown color is produced when peroxide oxidizes the normal red hemoglobin to dark brown methemoglobin—a material that cannot carry oxygen to the tissues. Before Takahara's discovery, it was believed that catalase in the blood was absolutely essential to life since hydrogen peroxide is produced in many metabolic reactions.

In fact, acatalasemia is a relatively mild disease when compared with many of the enzyme deficiency diseases. Any lesion in the mouth of an acatalasemic person provides suitable media for the growth of bacteria such as streptococcus and pneumococcus. These bacteria can split blood cells. They also produce hydrogen peroxide and they do not provide catalase. The peroxide oxidizes hemoglobin in the blood reaching the area, depriving it of its ability to carry oxygen. This leads to gangrene and the death of the tissue. The dead tissue fosters further bacterial growth, providing more peroxide to oxidize more hemoglobin. Thus, a vicious circle results. The circle must be broken by surgical removal of teeth and the dead and gangrenous tissues of the mouth. There are no other known pathological effects of the disease.

Dr. Robert Feinstein, a research biochemist at Argonne National Laboratories, developed a strain of acatalasemic mice for use in a cancer research project involving the effects of hydrogen peroxide. Such mice were killed by an intraperitoneal (into the abdomen) injection of hydrogen peroxide, although such an injection was not lethal to normal mice. A subcutaneous (under the skin) infusion of pure catalase was able to protect acatalasemic mice from the deadly effect of hydrogen peroxide. Such protection (a form of enzyme replacement therapy) lasted only as long as the active enzyme remained in the body. The catalase is recognized by the body as a foreign protein and it is destroyed and excreted. Repeated injection also causes antibody formation.

Catalase-Containing Artificial Cells

Dr. Thomas Chang of McGill University used the acatalasemic mice to demonstrate the usefulness of his artificial cells in treating enzyme deficiency diseases. Experimental animals included normal mice, untreated acatalasemic mice, and acatalasemic mice protected by injection of catalase solution or by intraperitoneal injection of artificial cells, which consisted of microencapsulated catalase. Animals from each of these four groups were injected with sodium perborate—a material which is decomposed by catalase and which produces visible pathological effects if not decomposed. The untreated acatalasemic mice became progressively less mobile, limp, and showed signs of labored breathing after the perborate injection. The enzyme deficient mice treated with artificial cells before the perborate injection showed slight immobility in the first 5 minutes after injection, but then became more active and started to move about freely after 20 minutes. Normal mice and acatalasemic mice treated with catalase solution recovered quickly after perborate injection. Thus, both catalase solution and catalase artificial cells could protect mice deficient in catalase from the effect of perborate. Twenty minutes after injection into the abdominal cavity, free catalase can be found in the blood and it is then rapidly destroyed as a foreign protein. Microencapsulated catalase remains in the abdominal cavity and retains activity for a much longer time, thus providing long-lasting protection.

Another approach involved the direct application of catalase or artificial cells to oral lesions. It was reasoned that, since the disease was discovered as a result of the application of peroxide to such lesions, this would be a good way to study therapeutic activity. Direct application of catalase solution might result in absorption of the enzyme and bring about immunological reactions after repeated use. Certainly, the free enzyme would not stay at the site of the lesion for a very long time. Microencapsulated catalase, on the other hand, would be more stable than the free enzyme, would cause no immunological reactions, and could be kept at the site of the lesion. Acatalasemic mice were anesthesized and oral lesions were created surgically. When hydrogen peroxide was applied to these lesions, no foaming was seen and a brown color appeared at the site of peroxide application. Enzyme deficient mice, treated with a paste of microencapsulated catalase at the site of the lesion, produced enough enzyme to cause observable foaming. Encouraged by these results, Chang applied the microencapsulated catalase paste to different areas of the gingiva of a normal human subject (apparently himself). No local irritation or other observable adverse effects were noted. The microencapsulated catalase efficiently destroyed peroxide applied to the gingiva.

Careful oral hygiene can usually prevent manifestation of symptoms of acatalasemia, but, if oral lesions do form, dental surgery and bone grafting in the affected region are usually required. Research, such as that carried out by Dr. Chang, makes available a much gentler and less traumatic form of therapy.

Acatalasemia involves a deficiency of an enzyme circulating in the blood. The enzyme is one that decomposes a very active material—hydrogen peroxide. In the absence of catalase, hydrogen peroxide is decomposed in its pathological reaction with hemoglobin.

More commonly, the deficient enzyme is not normally found in the circulatory system, but rather in an organ such as kidneys, liver, or spleen. The substances decomposed by these enzymes are not as chemically reactive as hydrogen peroxide and, in the absence of the enzyme, they accumulate in the tissues. This accumulation produces the pathological effects seen in these "storage diseases."

Among the most interesting and instructive of these maladies are those involving the storage of amino acids, glycogen, and lipids (fatty materials). Consider first a disease of amino acid metabolism—phenylketonuria, frequently referred to as PKU.

Phenylketonuria

This is the most severe of four diseases known to involve deficiencies of enzymes of metabolism of the essential amino acid, phenylalanine. The first step in this metabolism is the conversion of phenylalanine to tyrosine by the enzyme, phenylalanine hydroxylase, shown in Equation 13–1. This enzyme is missing in PKU victims.

It has been estimated that 1 of every 25,000 persons descended from Northern European ancestors has PKU. Most of the patients are found in mental institutions where they comprise about 1% of the mentally defective population. Most PKU victims have I.Q.s below 20; few are above 50. Many are unable to walk, talk, or control excretory functions. The most frequent reasons given for committing these people to institutions involve behavior problems. The "high-grade" patient is de-

Equation 13–1

$$\text{Phenylalanine} \xrightarrow[\text{(Phenylalanine hydroxylase)}]{} \text{Tyrosine}$$

scribed as shy, anxious, and restless, while more serious cases exhibit noisy, destructive psychotic episodes. Hyperactivity, irritability, and an uncontrollable temper are frequent characteristics.

A deficiency of phenylalanine hydroxylase causes excess phenylalanine to accumulate in the blood and spinal fluid. All important pathological changes observed in this disease involve the central nervous system including the brain. Indeed, this organ has only two-thirds of the weight of a normal brain in many cases. The formation of myelin, a unique membrane material which protects the cells of the central nervous system, is defective. Abnormal electroencephalogram patterns are seen in about three-fourths of phenylketonurics.

Phenylalanine hydroxylase normally appears in the liver during biochemical differentiation that occurs after birth. A deficiency of this enzyme is revealed at this time by an elevation of plasma phenylalanine levels to as much as 30 times normal and by the excretion of phenylpyruvic acid in the urine. The test for this latter material is extremely simple: several drops of 5% ferric chloride solution are added to fresh urine. A solid forms due to precipitation of phosphates. As more ferric chloride is added, there appears in positive specimens after 2 or 3 minutes an olive green color that fades in 1 to 2 hours. If this test is positive, the plasma phenylalanine level should be checked to confirm the diagnosis. Several states require that infants be tested for PKU before they leave the hospital.

It is essential that phenylketonuria be detected as early as possible and that treatment be started as soon as the diagnosis is confirmed. Treatment consists of a diet restricted in phenylalanine. Since this is an essential amino acid, its total elimination from the diet would be lethal.

The phenylalanine level must be carefully controlled and the patient should be checked frequently by a physician. The strictness of the diet can usually be relaxed by age 4.

The January 19, 1962, issue of *Life* magazine graphically illustrated the importance of early detection and treatment of this disease. Two sisters, ages 7 and 5, were pictured. Both had phenylketonuria. The older girl was diagnosed about a year after birth. Treatment was too late. She is in an institution for the mentally retarded. Her younger sister was diagnosed soon after birth and placed on a diet low in phenylalanine—a diet which maintained her mental and physical health.

Glycogen Storage Diseases

Glycogen is the reserve material of carbohydrate metabolism. It is a huge polymer made up of thousands of glucose molecules. These are linked together in a combination of linear and branched chains giving the molecule a treelike appearance. It is most abundant in the liver, normally comprising 5% of the weight of this organ. In muscle, 1% of tissue weight is glycogen. Other tissues have lesser amounts.

The late Nobel laureate, Dr. Gerty Cori, categorized six types of glycogen storage diseases during the 1950s. Today at least 10 are known —each due to a deficiency of a different enzyme. The most severe is Cori's type 2—Pompe's disease.

Except for the accumulation of glycogen, carbohydrate metabolism seems to be normal in Pompe victims. The infant patient is usually born free of symptoms, but has a deficiency of alpha-glucosidase. Glycogen is deposited, progressively leading to disruption of muscle fibers— including those of the heart. The disease manifests itself usually between the second and sixth months of life. Infants eat poorly, become listless, and fail to grow. There is vomiting, muscular weakness, drooling, labored breathing, and blueness in the face. Sometimes there is the appearance of imbecility like that seen in cretins and mongoloids. There may be nerve defects. Enlargement of the heart can usually be detected and there is frequently a heart murmur. Infants afflicted with this disease die of heart failure and in some cases bronchopneumonia before reaching their second birthday. At autopsy, the heart may weigh two to six times as much as a normal heart because of glycogen accumulation.

Pompe's disease is still considered a severe and incurable malady, but the advent of the age of enzyme replacement therapy offers some hope. The microorganism Aspergillus niger is a known source of alpha-glucosidase. Dr. George Hug and co-workers at the University of

Cincinnati Medical School infused an extract from this organism into a Pompe infant and produced some encouraging results. Muscle tone seemed to improve. Chest X-rays showed no further heart enlargement during therapy and the electrocardiogram improved. The concentration of glycogen in the liver was reduced to well within the normal range and the enzyme was recovered in the liver. Glycogen stored in muscle tissue and in the cerebral cortex appeared unchanged. After initial clinical improvement, Pompe symptoms worsened relentlessly for the last 6 weeks of treatment. No adverse reactions to treatment were seen and there were no immunologic complications. The patient died about 4 months after the start of therapy. At autopsy, the liver glycogen level was normal and the glycogen in the heart and skeletal muscle was no more than half that seen in autopsies of untreated Pompe victims.

Human placental tissue is a good source of the enzyme missing in Pompe patients and it is hoped that large amounts of the pure human enzyme will produce better results than the microbial extract.

Lipid Storage Diseases

The first recognition of lipid storage disease syndromes occurred around the end of the nineteenth century. There are now 10 of these diseases known. Although the diseases are caused by deficiencies of different enzymes and produce different symptoms, the lipids (fatty materials) which accumulate are closely related. All contain a material called ceramide which has the structure:

$$CH_3-(CH_2)_{12}-CH=CH-CH(OH)-CH(NHR)CH_2-O-X.$$

The materials differ in having different groups attached to the end of the molecule designated X.

Therapy for these diseases has been largely supportive. That is, the goal has been to relieve pain and make the patient as comfortable as possible. There was little progress in therapeutic methods in the 65 years immediately following Garrod's definition of "inborn errors of metabolism." A new era in the history of the treatment of enzyme deficiency diseases started around 1970—the era of enzyme replacement therapy.

To a reader familiar with insulin replacement therapy used in diabetes, enzyme replacement therapy may seem to be a straightforward procedure. Unfortunately, it is not. Insulin is a protein hormone, which acts in the bloodstream. The enzymes in question are normally found in vital organs, in subcellular particles called "lysosomes." Ideally, the replacement enzyme should leave the bloodstream and be taken up by the lysosomes of the organ which stores the lipid. Clinical research must determine how close we are to seeing the fulfillment of this ideal. If pure human enzymes cannot reach the target area, it is possible that attachment of carrier molecules will direct the enzyme to the desired site.

To illustrate some of the techniques of enzyme replacement attempted to date, we shall consider three lipid storage diseases: Tay-Sachs, Fabry, and Gaucher diseases.

Tay-Sachs Disease

Tay, an opthalmologist, recognized inherited changes in the fundus (part of the eye opposite the pupil) in certain children. Sachs, a neurologist, found that these children had severely impaired cerebral development. It is now known that there are various forms of Tay-Sachs disease, all of which are due to the accumulation of fatty materials in the brains of affected children. This accumulation is due to the absence of one or more forms (isoenzymes) of the enzyme, hexosaminidase. The disease attacks Jewish infants most frequently and it leads to blindness, seizures, neurologic deterioration, and death usually before the age of 3. No therapy has yet been found successful.

A recent report from the National Institutes of Health described attempted enzyme replacement therapy involving a 12-month-old girl showing classic Tay-Sachs symptoms and diminished levels of the two forms of hexosaminidase in serum. The enzyme was purified 6,000-fold from human urine and injected into the patient. The serum level of the enzyme rose to normal but it fell rapidly and reached preinjection levels in 75 minutes. There was a decrease in the plasma level of the lipid substrate and a corresponding increase in the concentration of decomposition products. There was a significant increase of the enzyme concentration in the liver, but no enzyme seemed to reach the brain or central nervous system. The clinical course was neither improved nor worsened by the enzyme treatment. The patient died 7 months later of Tay-Sachs disease.

The failure of the enzyme to reach the brain or central nervous system points out another problem in treating diseases attacking those areas. Injection of enzyme directly into the central nervous system would cause severe fever. Therefore, the enzyme is put into the bloodstream. The so-called blood-brain barrier[1] makes it very difficult to transfer ma-

---

[1]The blood-brain barrier results from the extreme impermeability of capillaries in the brain. The reason for this low permeability is unknown.

terials from the blood into the brain or central nervous system. The problem is somewhat akin to the problem of mixing oil and water, since blood tends to dissolve ionic (electrically charged) materials while the brain has a higher level of fatty materials. Methods are being sought for temporarily opening the blood-brain barrier. Animal studies have shown that present methods are not safe enough for human trials. Successful enzyme replacement therapy for diseases attacking the central nervous system and the brain will probably follow the development of safe methods for crossing the blood-brain barrier. The use of carrier molecules may well provide the key.

Gaucher's Disease

There are basically two forms of Gaucher's disease—an infantile form and an adult form. The former is the more severe form. The infant is usually normal at birth, but soon becomes apathetic. There is progressive physical and mental retardation. The abdomen increases in size with spleen and liver enlargement. These organs, as well as the lymph nodes and bone marrow, contain distorted cells called "Gaucher cells." Nerve defects appear with low muscle tone and the infant assumes an arched-back position. There is difficulty in swallowing, severe coughing, and intermittent bluish skin. Gaucher cells invade the lungs. Progressive wasting and weakness result in death usually before the first birthday.

The more common adult form becomes evident at any age. Patients show anemia and suffer the effects of an overly active spleen. Bone defects are common. There is patchy brown pigmentation of the skin, especially over the fronts of the legs and in the cheek region. The outcome of the disease is variable but many patients lead long, nearly unaffected lives, especially if the spleen is removed.

Gaucher's disease was the first lipid storage disease in which the metabolic derangement was demonstrated conclusively. The accumulating lipid in Gaucher's disease is glucocerebroside—a molecule having a glucose residue attached at the end of the ceramide chain. Brady and co-workers at the National Institutes of Health synthesized glucocerebroside with radioactive carbon in the glucose part. Using this "labeled" substrate, they found in all tissues of the body an enzyme which breaks down glucocerebroside.

$$\text{Glucocerebroside and Water} \xrightarrow{\text{Enzyme}} \text{Glucose and Ceramide}$$

There was high activity of this enzyme in the spleen and it was purified from this tissue. The activity of the enzyme was measured in samples of human spleen tissue taken at operations from a control series of patients and from Gaucher's disease patients. Those having adult Gaucher's disease had approximately 15% of the activity seen in controls, while those having the infantile form had extremely little, if any, glucocerebrosidase activity. The use of radioactive labeling of substrates has been used to study other enzyme deficiency diseases.

Because the normal spleen is the site of the highest concentration of glucocerebrosidase, a spleen transplant was attempted on a patient with terminal juvenile Gaucher's disease. The graft functioned for 40 days and then failed due to rejection. The patient died from the original disease 3 months after the operation. Rejection is a serious problem in spleen transplantation and such therapy is not likely to succeed until the antigenicity of the spleen can be overcome.

Brady's group at the National Institutes of Health has isolated and purified glucocerebrosidase from human placental tissue. They have treated two patients with this enzyme and have reported definite decreases in the quantity of accumulated lipids in one patient suffering from adult Gaucher's disease and one having the juvenile form.

Fabry's Disease

Observation of a characteristic skin lesion permitted the dermatologist Fabry to describe the disease that bears his name. This malady results from the generalized visceral deposition of a lipid called cerebroside trihexosyl (CTH). The material accumulated in the peripheral nervous system causes excruciating bouts of pain and acroparesthesias—a disease marked by attacks of tingling, numbness, and stiffness in the extremities, especially fingers, hands, and forearms. The lipid deposits obstruct blood vessels depriving tissues of blood. Fabry's patients are often unable to perspire. Death usually results in adulthood from renal (kidney) failure, from diseases of the circulation, or from the effects of high blood pressure on the heart or brain. The enzyme deficient in this disease, ceramide trihexosidase, is a specific galactosidase which cleaves a single galactose residue from the end of the CTH molecule.

The age of enzyme replacement therapy for Fabry's disease started in 1970 although pure enzyme was not available for clinical use at that time. Plasma containing normal levels of ceramide trihexosidase activity was infused into two Fabry patients. The enzyme could still be detected in the patient's bloodstream 7 days after infusion. A sharp reduction in CTH was seen 6 to 18 hours after plasma infusion—a time coincident with maximum enzyme activity. An unexpected enhancement of CTHase activity in these patients was reported. The activity found in the blood was 150% of that injected, suggesting that the infused plasma may have activated the patient's own inactive enzyme.

This mode of therapy suffers from the requirement of frequent infusion of large quantities of plasma. This taxes already defective kidneys and may ultimately provoke immunological problems. The method has been criticized by several groups who were unable to duplicate the results.

The need for frequent infusion of high quantities of plasma prompted a search for a continuous source of active enzyme. Since an important site of CTH accumulation is the defective Fabry kidney and since normal renal tissue contains substantial ceramide trihexosidase activity, kidney transplantation was attempted. The technique has had mixed success. Patients helped by the transplant have reported an end to periodic bouts of excruciating Fabry pain and an increased ability to perspire. They have described themselves as free of symptoms. Attending physicians have noted marked clinical improvement and a significantly greater level of enzyme activity in the plasma.

The use of kidney transplantation in treating Fabry's disease has been criticized in a recent article in the *Journal of the American Medical Association*. It was pointed out that, although Fabry's disease did not recur in any transplanted kidney, only three Fabry's disease patients of 11, who received such transplants, have functioning grafts 1 year after transplantation. No alternative mode of therapy was suggested.

Brady's group at the National Institutes of Health has provided an alternative therapeutic method—direct enzyme replacement. Ceramide trihexosidase was highly purified from human placental tissue and given intravenously to two Fabry patients. It was rapidly cleared from the blood and taken up mostly by the liver. In both patients, there was a significant decrease in the level of CTH circulating in the blood. Infusion of blood platelets and leukocytes suspended in plasma had effects similar to those produced by the enzyme injection. This therapy caused CTHase activity in serum to rise 68% over preinjection levels.

The age of enzyme replacement therapy is still in infancy. There have been disappointments, but there is much reason for hope. Before the advent of such therapy, diagnosis of many of the enzyme deficiency diseases was tantamount to a death sentence. Although many victims of the diseases still die from them, for the first time in the history of medical science, there is available a form of therapy which may remove the label of "always fatal" from the diseases. Whether the enzyme is replaced by organ transplantation, by injection of purified human enzyme, or by some other method, the era of enzyme replacement therapy has arrived and it is an era of great hope and promise.

Addendum: Liposomes are highly ordered aggregates that form when phospholipids (combinations of phosphoric acid with fatty materials) are confronted with water. Recent studies by Gregoriadis and others have shown that these liposomes may be used to direct enzymes or drugs to certain organs. When injected into the veins of animals, the liposome-enzyme complex is taken up largely by the liver and the spleen and, to a lesser extent, by the kidneys, lungs, skeletal muscle and brain. Evidence obtained by the use of biochemical techniques and electron microscopy indicate that the enzymes are localized within the lysosomes (sub-cellular particles) within these organs, exactly where they are needed.

It seems appropriate to mention, in closing this chapter, that a good deal of research on these enzyme deficiency diseases is funded by the National Foundation–March of Dimes.

Chapter Fourteen

Synthesis of Enzymes and Enzymes in Synthesis

WHEN I TOOK my first biochemistry course in 1958–1959, the instructor said, "I used to say that they'll never synthesize an enzyme in the laboratory; now I say that they'll probably never synthesize an enzyme in the laboratory." In January, 1969, two research groups, working independently and using entirely different methods, described the synthesis of an enzymically active protein. Both groups had synthesized ribonuclease.

Why is this such a big deal? Why did the instructor predict that it would never be done? The answer to these questions is found in the complexity of the protein structure and in the fact that only the L–form amino acids are found in enzymes.

As we have seen in Chapter 1, there are 21 amino acids commonly found in proteins. These amino acids can occur in any proportions and in any sequence in a given protein. But the proper proportions and the proper sequence are needed for enzymic activity.

Thus, the first job is the determination of the amino acid composition of the protein. This is a multistep process. An accurately weighed sample of the protein is broken down to its constituent amino acids by subjecting it to hydrochloric acid in an evacuated sealed tube at 110° C for 12 to 96 hours. The products from this acid hydrolysis are separated using a physical technique known as ion-exchange chromatography. Quantitative analysis is then used to determine how much of each amino acid there was in the protein, by measuring the amount of each liberated by the acid.

Tryptophan is completely destroyed by this acid hydrolysis. To determine the amount of this amino acid present in the protein, it is necessary to decompose a second carefully weighed sample of the protein using sodium hydroxide solution at 100° C for 4 to 8 hours. This procedure destroys cystine, cysteine, serine, threonine, and arginine, but leaves tryptophan intact. The products are separated by the chromatographic method and tryptophan is determined quantitatively.

When the amino acid composition is established, it is necessary to determine the sequence in which these building blocks occur in the protein. Many proteins consist of two or more polypeptide chains. It is necessary to determine how many chains there are, and to isolate the individual chains. They are then cleaved to give smaller chains of more manageable size. The amino acid sequence of these fragments must be determined in a way which gives information on how the fragments fit together in the enzyme. Finally, it is necessary to determine how, and at what amino acid sites, the chains are held together.

The molecular weight of the protein is usually determined by use of a device known as an ultracentrifuge. This device separates molecules on the basis of their molecular weights by exposing them to intense gravitational forces. It provides a sensitive method for determining the molecular weights of large molecules.

When all of this information is available, the number and sequence of amino acid residues are known and the "primary structure" of the protein is established. Although there are complex interactions in protein structures, which are referred to as secondary, tertiary, and quaternary structures, it is not necessary to know these to carry out the synthesis. When the primary structure—the proper sequence of amino acids—is prepared, the molecule tends to arrange itself in the proper conformation to give the correct interactions.

One is now ready to devise a method for bonding all of the amino acids in the proper sequence without destroying the optical activity of any of the amino acids. Only L–forms can be included in the synthesis and the amino acids must remain in the L–form in the protein.

It is remarkable that two groups, working independently and using different methods, accomplished this feat simultaneously. Denkewalter and his group at the Merck Sharp and Dohme Research Laboratories synthesized a form of ribonuclease using materials in solution. Merrifield and co-workers at the Rockefeller University used "solid-phase synthesis," in which the terminal amino acid is attached to an insoluble support and amino acids are added one by one to form the protein molecule. A slightly different form of ribonuclease was prepared by this method.

Ribonuclease is probably the most widely studied enzyme from many different areas of interest. Its primary structure had been determined in 1963 by Smyth, Stein, and Moore at the Rockefeller Institute (later to become the Rockefeller University). It is a small enzyme having a molecular weight just under 14,000. It contains 124 amino acid residues in a single chain. The molecule has all of the usual amino acids with the exception of tryptophan. Ribonuclease can be selectively split by the enzyme subtilisin, giving two inactive chains—one containing 20 amino acid residues, the other 104. The former chain is called S–peptide; the latter, S–protein. When the two are mixed in solution in equivalent amounts, enzymic activity is restored, even though they do not combine chemically.

The Denkewalter group at Merck synthesized the S–protein in solution and mixed it with natural S–peptide to generate ribonuclease activity. The Merrifield group at Rockefeller University synthesized the total enzyme using the solid-phase approach.

While it is not difficult to form the peptide bond, the chore of forming the bonds linking the proper amino acids in the proper sequence with retention of optical activity is an imposing one. If 124 amino acid molecules containing proper amounts of each of the 20 amino acids of ribonuclease were simply reacted at random, there would be 20^{124} possible arrangements (20^{124} is 20 multiplied by itself 123 times).

In order to accomplish the required order, it is necessary to bond the amino acids together one by one. After each reaction, there is a free amino group at one end of the chain and a free carboxyl group at the other end. These are called the amino terminal amino acid and the carboxyl terminal amino acid, respectively. The amino group of the acid which is to be the amino terminal amino acid is "protected" by use of a "blocking reagent" before the coupling reaction is run. If the amino group were not protected, another molecule of the same amino acid could add to it after the carboxyl group was activated and before the second reactant was mixed with it.

The carboxyl group of the protected amino acid is activated and the carboxyl terminal amino acid (or peptide) is mixed with it. A peptide bond results. The blocking reagent is removed from the terminal amino group and one step of the synthesis has been completed. This sequence of steps is repeated for each amino acid added to the chain. The blocking groups must remain in place during peptide bond formation and must be easily removable without breaking any other bonds in the growing peptide chain. In addition to the amino functions, the SH groups of cysteine must be protected by blocking groups.

The Merck group used N–carboxyanhydrides and activated esters to produce polypeptides and joined these fragments together by use of an azide reaction to get the desired S–protein. (N–carboxyanhydrides are doubly useful, since this group simultaneously blocks the amino function and activates the carboxyl group.) Examples of the N–carboxyanhydride, activated ester, and azide reactions are shown in Figure 14–1.

SOLID-PHASE SYNTHESIS

All of the reactions carried out in the Merck synthesis were done in solution. Merrifield devised a method of synthesis in which the amino acids were removed from solution as the result of their reaction.

The synthesis starts with a solid support—polystyrene cross-linked with 1% to 2% divinylbenzene and substituted with chloromethyl groups. This support is insoluble in all solvents used, but it swells when exposed to solvent. Swelling is important since it permits reactions to occur inside the pores as well as on the surface. The pores are extremely large and easily accommodate the product proteins.

The first amino-protected amino acid is dissolved and the support is contacted with the solution. The amino acid bonds chemically to the support. The blocking group is selectively removed, usually by trifluoracetic acid, and the next amino-protected amino acid is activated and added. The deprotection and coupling steps are repeated until the protein chain is completed. The chain is then cleaved from the support without breaking any of the peptide bonds of the protein. Hydrofluoric acid was used to accomplish this in the case of the ribonuclease synthesis.

The amino-protecting group used in the synthesis was the tertiary butyloxycarbonyl (t-Boc) group. The most frequently used activating reagent for the carboxyl groups was dicyclohexyl carbodiimide. Some of the important reactions of the solid phase method are illustrated in Figure 14–2.

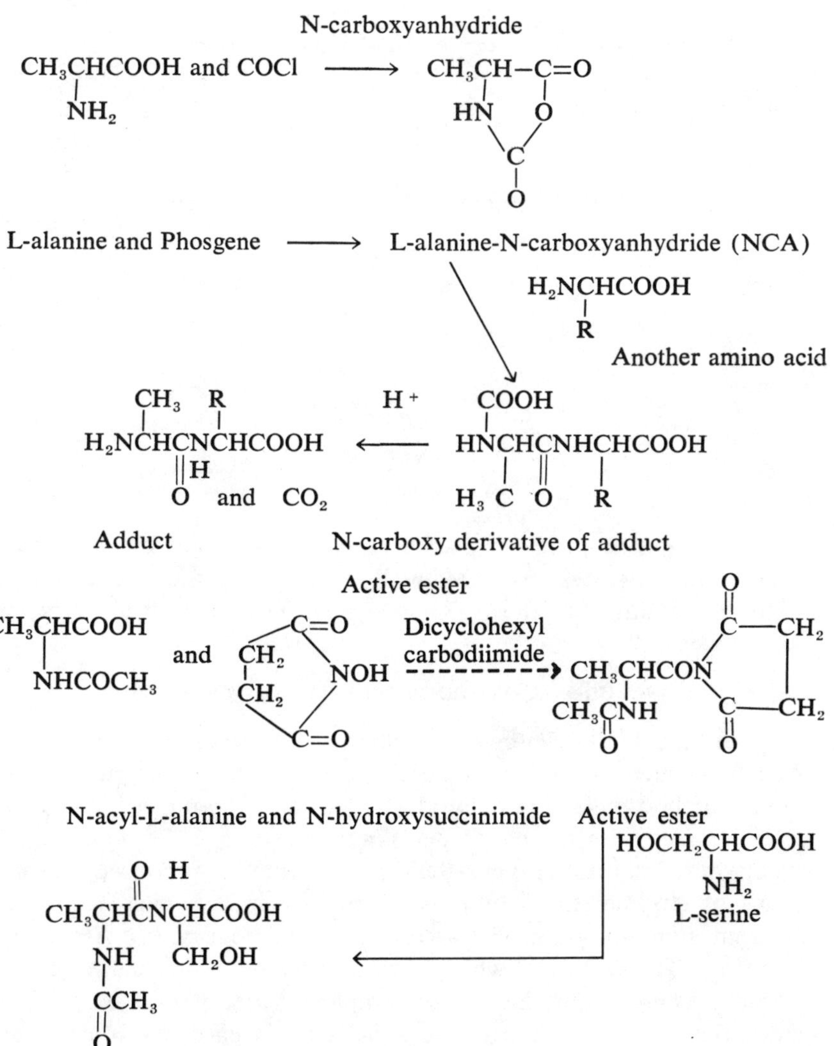

Figure 14–1   Some reactions used in synthesis of enzymes.

183

Figure 14–1   continued

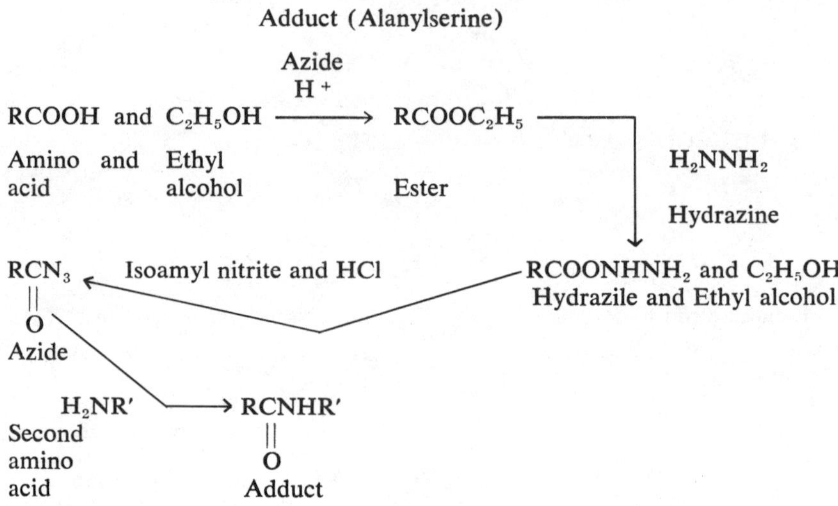

Solid-phase synthesis has been automated and a considerable saving of time has resulted. "Manual" syntheses can add two or three amino acid residues to the chain in a day. The automatic synthesizer adds six per day.

CONCLUDING COMMENTS ON ENZYME SYNTHESIS

Although the synthesis of ribonuclease represents an epic advance in science, the methods are still in a rather rudimentary stage. Neither method gave a pure product and neither produced much material. The activities of the products were between 14% and 20% of that of "pure" natural enzyme. Solid-phase synthesis produced 100 milligrams of enzyme; the solution method produced 5 micrograms. (A milligram is one-thousandth of a gram and a microgram is one-millionth of a gram.) The syntheses, which were possible only after many years of research, required many months to complete. Pure, natural enzymes are synthesized in the cell in less than 3 minutes. Science has come a long way, but still has a long way to go.

SOME USES OF ENZYMES IN ORGANIC SYNTHESIS

The synthesis of an enzyme provides examples of reactions that occur under very mild conditions. But mild conditions are more the ex-

Figure 14–2   Some reactions used in solid-phase synthesis.

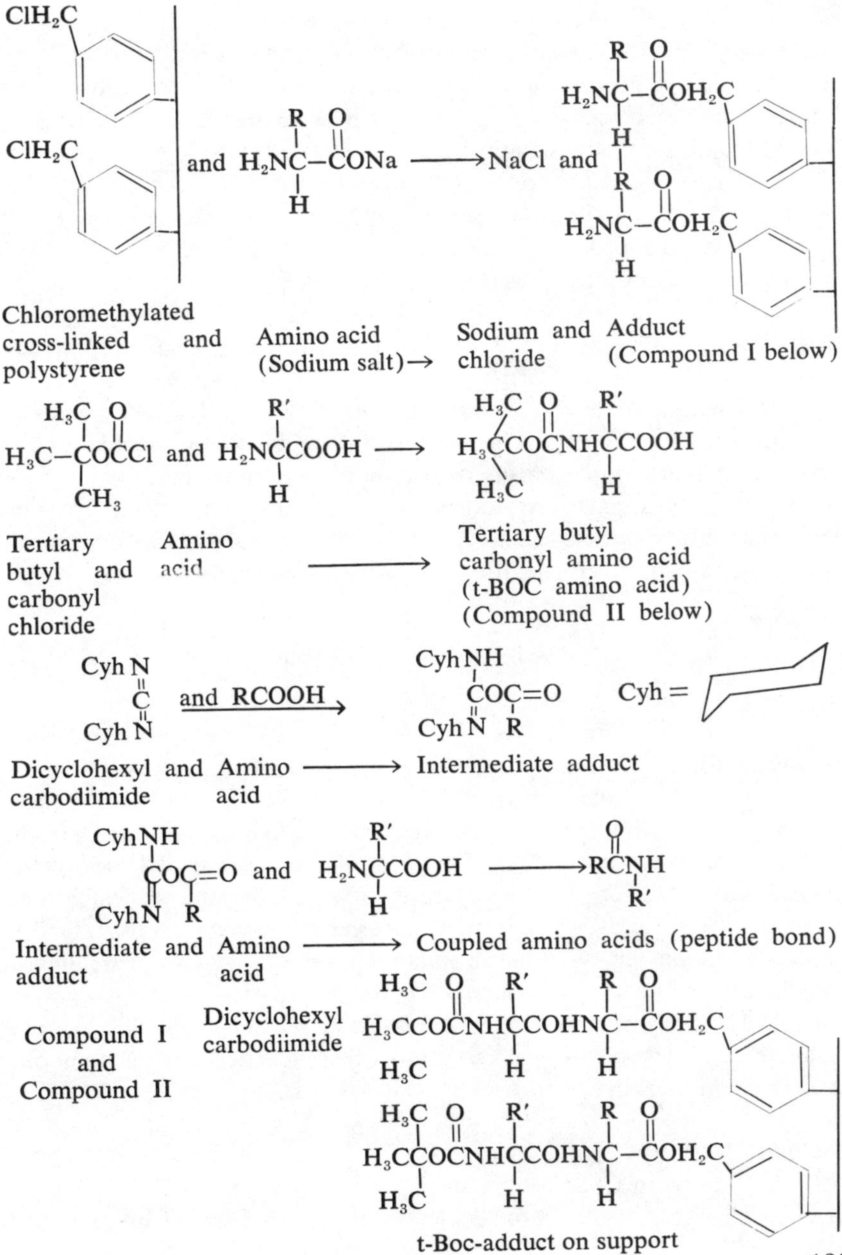

ClH_2C— … and $H_2NC(R)(H)\text{–}CONa$ (with $C=O$) $\longrightarrow$ NaCl and $H_2NC(R)(H)\text{–}COH_2C$— … (Compound I below)

Chloromethylated cross-linked polystyrene   and   Amino acid (Sodium salt) $\rightarrow$ Sodium chloride and Adduct (Compound I below)

$H_3C\text{–}COCCl$ [with $(CH_3)_2$ and $C=O$] and $H_2NCCOOH$ [with R' and H] $\longrightarrow$ $H_3CCOCNHCCOOH$ [with $(H_3C)_3$, $C=O$, R', H]

Tertiary butyl carbonyl chloride   and   Amino acid $\longrightarrow$ Tertiary butyl carbonyl amino acid (t-BOC amino acid) (Compound II below)

Cyh N=C=N Cyh   and   RCOOH $\longrightarrow$ Cyh NH–COC=O / Cyh N=R   ;   Cyh = (cyclohexyl)

Dicyclohexyl carbodiimide   and   Amino acid $\longrightarrow$ Intermediate adduct

Cyh NH–COC=O / Cyh N–R   and $H_2NCCOOH$ [with R' and H] $\longrightarrow$ $RCNH$ [with O and R']

Intermediate adduct   and   Amino acid $\longrightarrow$ Coupled amino acids (peptide bond)

Compound I and Compound II   Dicyclohexyl carbodiimide $H_3CCOCNHCCOHNC\text{–}COH_2C$— … [with H_3C, O, R', R, O, H_3C, H, H]

$H_3CCOCNHCCOHNC\text{–}COH_2C$— … [with H_3C, O, R', R, O, H_3C, H, H]

t-Boc-adduct on support

185

ception than the rule in organic synthesis. More generally, the reactions are carried out in boiling solvents using high temperatures and strong acids or bases. Sometimes high pressures are required. It is the use of these harsh conditions which leads to low yields, mixed products, and tar formation when organic reactions are carried out. Any reactions that can be completed without resort to harsh conditions should be eagerly sought by the synthetic organic chemist. Enzymes provide such reactions.

There are additional advantages of enzymes as synthetic reagents. Their specificity limits the extent of side reactions, so high yields of relatively pure products are obtained. Enzymes can effect conversion of one specific site in a molecule, leaving other sites intact. The stereospecificity of many enzymes makes them useful for preparing optically active compounds—a job that is very tedious when classical methods are used.

The instability of enzymes has doubtless retarded their exploitation as synthetic reagents. The advent of immobilization techniques has resulted in more stable enzymes, negating this drawback. We have already seen how Japanese scientists have exploited immobilized enzymes and immobilized whole cells for industrial syntheses. It seems appropriate, at this point, to mention a number of other uses of enzymes in synthesis.

Semisynthetic Penicillins

Penicillin amidase is able to cleave penicillin G, a common natural form of penicillin, to give 6–aminopenicillanic acid (6–APA). This is illustrated in Equation 14–1. This is a key reaction in the preparation of semisynthetic penicillins from natural penicillin G which is prepared by fermentation. Simple chemical reactions are then used to prepare the desired semisynthetic products. These products are intended to improve the clinical results of penicillin therapy. Some improvements which have been noted are activity against gram-negative bacteria, better stability and absorption characteristics, increased resistance to penicillinase-producing pathogens, and decreased allergenicity.

The enzyme reaction makes 6–APA readily available at a reasonable cost. A good deal of research is being carried out on immobilization of the enzyme to further reduce the cost.

Steroid Conversions

We saw in Chapter 4 that microorganisms are used to effect important changes in steroid molecules—changes that would be extremely

Equation 14-1

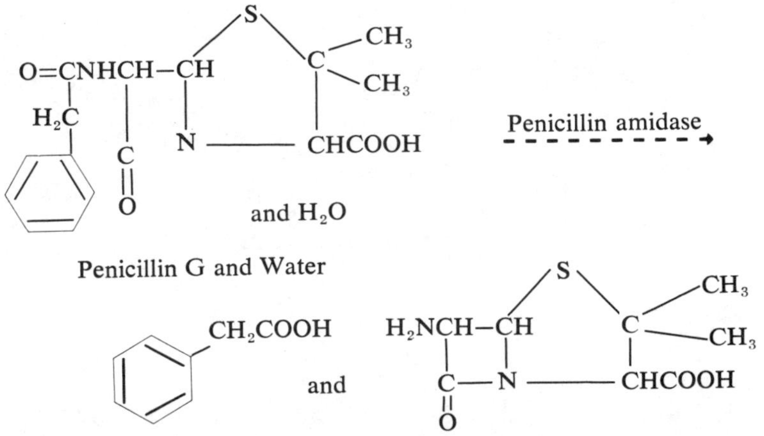

Penicillin G and Water

Penicillin amidase $\longrightarrow$

and H_2O

Phenylacetic acid and 6-aminopenicillanic acid (6-APA)

difficult using purely chemical means. The microorganisms, of course, function as a source of enzymes, which are the true catalysts for these conversions.

Another approach to these steroid conversions involves isolating the proper enzyme and using it in solution, or in the immobilized state, to catalyze the reaction of interest. An example of this is the conversion of cortisol to prednisolone by a steroid dehydrogenase isolated from the microorganism Corynebacterium simplex. The reaction is illustrated in Equation 14–2. The enzyme has been isolated and entrapped in a cross-linked polyacrylamide gel. The immobilized enzyme was an effective catalyst for the dehydrogenation.

RADIOLABELING WITH ENZYMES

An unusual application of enzymes was reported recently by Cohen and co-workers at the Los Angeles Veterans' Hospital. They prepared pharmaceutically pure amino acids labeled with radioactive nitrogen–13. This isotope is useful for labeling many small biologically active compounds where an externally detectable label is required. The preparation of the labeled compounds must be completed very quickly as the half-life of the radioactive nitrogen is only 10 minutes (half of the

Equation 14–2

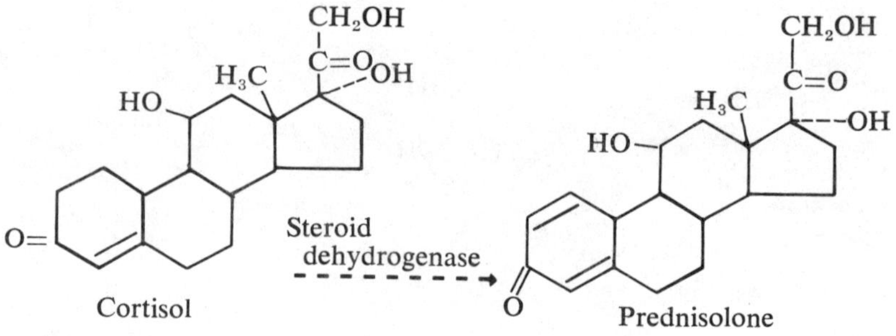

$$CH_2OH$$

Cortisol

Steroid
dehydrogenase

Prednisolone

material decomposes in 10 minutes). Chemical syntheses are relatively slow and they give impure products. Compounds having optical isomers are produced as racemic mixtures (equal quantities of D– and L–isomers). Enzymic synthesis is rapid and it yields relatively pure products, including optically active compounds.

But enzymes create problems in obtaining pharmaceutically pure materials. The enzyme itself contaminates the product bringing the possibility of antigenic and pyrogenic reactions. This difficulty is avoided by the use of immobilized enzymes.

Cohen's group bonded glutamic dehydrogenase and glutamic-pyruvic transaminase separately to silica beads. Ammonia containing [13]N (radioactive nitrogen–13) was obtained from the cyclotron at the University of California at Los Angeles (a device used to prepare radioactive isotopes). This was mixed with alpha-ketoglutarate and NADH and the mixture was passed through a column of the immobilized glutamic dehydrogenase. The product is ^{13}N–L–glutamic acid. This can be mixed with pyruvic acid and passed through a second column containing the immobilized transaminase. In this reaction, ^{13}N–L–alanine is formed. The reactions in this sequence are shown in Equations 14–3 and 14–4.

Previous examples have indicated ways in which the use of immobilized enzymes has made processes more convenient or more economical. This provides an instance in which the use of immobilized enzymes makes a process possible.

But it should not be assumed that an enzyme must be immobi-

Equation 14–3

$$NH_4^{\oplus} \text{ and } NADH \text{ and } \begin{array}{c} COOH \\ | \\ C=O \\ | \\ CH_2 \\ | \\ CH_2 \\ | \\ COOH \end{array} \xrightarrow[\text{dehydrogenase}]{\text{Glutamic}} \begin{array}{c} COOH \\ | \\ HCN^{13}H_2 \\ | \\ CH_2 \\ | \\ CH_2 \\ | \\ COOH \end{array} \text{ and } NAD \text{ and } H_2O$$

Ammonium ion and Cofactor
and Alpha-ketoglutarate

$$\xrightarrow{\text{Enzyme}} {}^{13}\text{N-L-glutamic acid}$$
and Cofactor and Water

Equation 14–4

$$\begin{array}{c} COOH \\ | \\ HCN^{13}H_2 \\ | \\ CH_2 \\ | \\ CH_2 \\ | \\ COOH \end{array} \text{ and } \begin{array}{c} CH_3 \\ | \\ C=O \\ | \\ COOH \end{array} \xrightarrow[\text{transaminase}]{\substack{\text{Glutamic} \\ \text{pyruvic}}} \begin{array}{c} COOH \\ | \\ C=O \\ | \\ CH_2 \\ | \\ CH_2 \\ | \\ COOH \end{array} \text{ and } \begin{array}{c} CH_3 \\ | \\ HCN^{13}H_2 \\ | \\ COOH \end{array}$$

$$\xrightarrow{\text{Enzyme}}$$

^{13}N-L-glutamic acid and Pyruvic acid $\longrightarrow$ Alpha-ketoglutaric acid
and ^{13}N-L-alanine

lized to be useful in synthesis. Sometimes a crude enzyme preparation
can be used and it may be cheap enough to permit its use on a one-time
batch basis. Recently, I completed an assignment at the University of
Nebraska in which I was asked to produce compounds known as cyclodextrins. These are prepared by treating starch with a microorganism
known as Bacillus mascerans, or with a specific enzyme obtained from
this bacterium. We were fortunate enough to obtain the enzyme free of

charge from the Corn Products Co. and the desired product was synthesized. There is no other known method of synthesis of cyclodextrins.

Many other examples of enzymic syntheses could be given, but this should suffice to show the potential usefulness of enzymes in organic synthesis.

Chapter Fifteen

A Potpourri of Enzyme Applications

BIOCHEMICAL FUEL CELLS

A FUEL CELL is a device which converts chemical energy into electrical energy. This same definition applies to the "dry cell," which powers flashlights and portable radios, as well as to the "wet cell," which starts cars. In the dry cell, the reactive chemical materials are eventually consumed and the "battery" is replaced; in the wet cell, the reactive chemical materials are continuously regenerated by the alternator or generator. In the fuel cell, the reactive chemical materials are continuously added to the cell and the cell generates electrical energy as long as this "fuel" is available.

A biochemical fuel cell is one which uses enzyme reactions to convert the fuel. The catalyst may be a free enzyme in solution, an immobilized enzyme, a microorganism, or an immobilized microorganism (whole cell).

The electrodes of the fuel cell are not changed in any way when the cell is operated. The capacity of the cell is governed by the size of the fuel reservoir, and the battery size is related to the rate of fuel conversion or power output.

A hydrogen-oxygen fuel cell was described by Grove in 1839. Seventy-two years later, Potter demonstrated that single-cell microorganisms, such as bacteria and yeasts, were able to generate electrical

191

currents by fermenting organic matter. Using various species and various substrates, he could generate 0.35 volt. Using enzymes alone, he got only 0.02 to 0.05 volt. In 1931, Cohen connected growing bacterial cultures in parallel and series and showed the ability of microorganisms to generate electrical power. He obtained a current of 2 milliamperes at a potential of 35 volts.

Quite often, a biochemical fuel cell consists of a normal oxygen (or air) cathode and an anode immersed in a solution containing the enzyme or microorganism and a substrate. A semipermeable membrane separates the two electrode compartments. The biochemical reaction generates electrons, which flow through the external circuit to the cathode. The cathode compartment supplies an oxidant, such as air, continuously. The spent fuel is replaced by fresh substrate in a continuous manner. As long as the enzyme remains active, electrical energy is generated.

Uses of Biochemical Fuel Cells

Three major uses have been proposed for biochemical fuel cells. In underdeveloped countries without abundant fossil fuels, a biofuel cell could be used to generate electricity from animal or vegetable matter. Such a cell should be cheap, long-lived, and "foolproof" in operation. A second application would be for military purposes. Finally, a biochemical fuel cell, based on human waste products, could be useful in space travel.

Advantages and Limitations
of Biochemical Fuel Cells

Probably the greatest advantage of the biochemical fuel cell over "conventional" fuel cells is the vast number of chemical reactants (substrates) that can be used as fuel. This includes essentially all organic compounds that can be oxidized by enzymes or microorganisms. Most organic materials fill this requirement. The fuel material does not have to be purified; in the conventional fuel cell, the use of nonpurified fuel would rapidly poison the catalyst. Microorganisms serve as self-regenerating catalysts and they work well on crude fuels. The lower cost of the fuel is an important advantage of the biochemical fuel cell.

The cell is limited by the physical and chemical restrictions imposed by the enzyme or microorganism. Thus, factors such as temperature and pH must be maintained within limits defined by the stability of the catalyst. Microorganisms may be active over a pH range between 0.5 and 10.5 and a temperature range of 5° to 70° C, but the range for optimum activity is more restrictive.

A major disadvantage of the biochemical fuel cell is the low power levels which have been produced by the device thus far. The development of the biochemical fuel cell is an interesting possibility in a world which is rapidly depleting its supply of fossil fuels, and long-term research is devoted to such development.

COFACTORS AND IMMOBILIZED COFACTORS

We have mentioned cofactors in various sections of the book without giving much information about them. But most enzymes require cofactors for activity. Most cofactors are relatively small molecules ranging in molecular weight from 200 to 800. The largest cofactor, co-enzyme B_{12}, has a molecular weight of 1,580. Most cofactors (coenzymes) are soluble materials that are used in solution; others are tightly bound to the protein. These are called "prosthetic groups." Some cofactors, such as pyridoxal phosphate, undergo only a transient change during an enzymic reaction. When the reaction is completed, the cofactor is restored to its original form. Such cofactors are "self-regenerating" and are needed in only catalytic (very small) amounts. More commonly, 1 molecule of cofactor is converted for each molecule of substrate converted and it is converted to a form which cannot be used again in the reaction. These cofactors must be added in amounts equivalent to the amount of substrate. Chemists call this a "stoichiometric amount."

Cofactor regeneration is of great concern to enzyme engineers, because it is the key to a great deal of further development in enzyme technology. Most of the enzyme applications mentioned in this book involve the use of enzymes which do not require cofactors. The development of an efficient and economical means of regenerating cofactors will make many additional enzymes available for industrial uses. Until such development is accomplished, the relatively high cost of cofactors will retard the development of enzyme technology.

A few cofactors can be regenerated by exposure to oxygen gas. This is a most desirable regenerating system because it is cheap and it produces only water and the cofactor. More frequently, a consumable substrate, other than oxygen, is required. Thus, the system becomes contaminated with the extraneous chemicals needed to regenerate the cofactor.

Methods used in research laboratories studying the regeneration problem have involved chemical, electrochemical, and enzymic reactions. No method has yet been found which is commercially useful.

The hydrolysis of adenosine triphosphate (ATP) is the fundamental driving force for a number of biochemical processes including muscle contraction, photosynthesis, bioluminescence, discharge of electrical organs, and the biosynthesis of proteins, nucleic acids, complex carbohydrates, and lipids. Two laboratory methods used for regeneration of this cofactor are of interest.

Gardner and Whitesides developed a method for regenerating ATP from AMP (adenosine monophosphate) by use of two immobilized enzymes and acetyl phosphate, which is produced chemically. They bonded adenylate kinase and acetate kinase to sepharose (an insoluble carrier) and used the combination in a reactor to produce 1 gram of ATP per hour. Equations 15–1 and 15–2 illustrate the reactions.

Equation 15–1

$$\text{AMP and ATP} \quad \xrightarrow{\text{Adenylate kinase}} \quad 2\ \text{ADP}$$

$$\begin{array}{ll}\text{Adenosine and} & \text{Adenosine} \\ \text{monophosphate} & \text{triphosphate}\end{array} \quad \xrightarrow{\text{Enzyme}} \quad \begin{array}{l}2\ \text{Adenosine} \\ \text{diphosphate}\end{array}$$

Equation 15–2

$$\text{2 ADP and 2 CH}_3\text{COOPOH} \quad \xrightarrow{\text{Acetate kinase}} \quad \text{2 ATP and 2 CH}_3\text{COO}^-$$

(with OH above and OH below the CH_3COOPOH group)

$$\begin{array}{ll}\text{Adenosine and} & \text{Acetyl} \\ \text{diphosphate} & \text{phosphate}\end{array} \quad \xrightarrow{\text{Enzyme}} \quad \begin{array}{ll}\text{Adenosine and} & \text{Acetate} \\ & \text{triphosphate}\end{array}$$

Archer and co-workers used the photosynthetic apparatus (chromatophores) isolated from a bacterium, Rhodospirillum rubum, to regenerate ATP from ADP (adenosine diphosphate), inorganic phosphate, and light. A continuous ultrafiltration reactor has been operated for up to 72 hours. The stability of the chromatophores seems to be the limiting factor in this system. If adenylate kinase is added to the reactor, AMP can also be regenerated to ATP, just as in the previous example.

In addition to cofactor regeneration, a method of cofactor retention is needed for the further development of enzyme technology. Since cofactors are relatively small, soluble molecules, they must be held in the reactor by artificial means. Semipermeable membranes are available which can hold these small molecules, but such membranes also prohibit movement of all but the smallest of substrates and products.

By bonding the cofactors to polymers, the size of the molecule can be increased to any size desired. Semipermeable membranes which retain only larger molecules can then be used.

It is generally seen that cofactors attached to insoluble supports are not useful in enzyme reactions. Larsson and Mosbach found that NAD bonded to Sepharose had only 0.2% of the coenzyme function of the soluble molecule. But cofactors immobilized by attachment to soluble supports have been prepared with good retention of activity. It seems likely that soluble supports will provide a satisfactory solution to the problem of coenzyme retention. Those supports which are soluble or insoluble, depending on the environment (see Chapter 3), may be useful in this regard.

A discussion of immobilized cofactors leads quite naturally to our next topic—affinity chromatography.

AFFINITY CHROMATOGRAPHY

Affinity chromatography is a technique which makes use of the specific reactions of certain proteins to affect their purification. The reactions of enzymes with competitive inhibitors, cofactors, substrate analogs, or substrates are used for enzyme purification. The method is based on the ability of the protein to bind a molecule (called a ligand) specifically and reversibly. A biospecific adsorbent is prepared by chemically bonding to an insoluble support a ligand which interacts specifically with the desired protein.

The protein is purified by passing a crude extract or solution containing the protein through a column containing the biospecific adsorbent. Molecules not interacting with the ligand will pass freely through the column. Molecules having affinity for the ligand will be retarded— the greater the affinity, the more the protein is retarded. Molecules that are tightly bound to the carrier are eluted by changing the pH, the salt concentration, or occasionally, the temperature. Sometimes soluble ligands are added to free the protein from the column. The affinity of the protein for the ligand must be great enough to retard the protein in its passage through the column, yet weak enough to permit its elution without denaturation.

The carrier molecule should have minimal adsorptive power and should form a loose, porous network that permits free passage of solutions with a good flow rate. The chemical structure of the carrier must permit convenient and extensive attachment of the ligand under rather mild conditions. This attachment must be stable during adsorption and elution.

The ligand must have an affinity for the protein of interest and it must have available chemical groups which permit it to be bonded firmly to a support molecule. The ligand may have a specific affinity for only one protein, or a small number of proteins, or it may have affinity for a large group of proteins. Cofactors are in the latter category.

Examples of this are provided by nicotine adenine dinucleotide (NAD) and nicotine adenine nucleotide phosphate (NADP). These cofactors may be used to isolate dehydrogenases from solution. They are probably the least specific of ligands.

Inhibitors and substrate analogs are generally much more specific—that is, they have an affinity for fewer enzymes than do the cofactors. A competitive inhibitor, p–amino–phenyl–beta–D–thiogalacto-pyranoside, was used by Steers and co-workers to purify lactase obtained from E. coli. This work is frequently cited in scientific publications because it revealed a new facet of the biospecific adsorbent. Steers found that the distance between the carrier and the inhibitor is a significant factor in determining the ability of the ligand to bind the protein. Thus, when the ligand was bonded directly to the support, no specific binding of lactase was observed.

A short "spacer arm"—a chain 10 angstrom units[1] long—inserted between the carrier and the ligand caused lactase to emerge from the column slightly behind the major protein mass. But a very long arm (about 21 angstroms long) between ligand and carrier produced a biospecific adsorbent that adsorbed the lactase very strongly. The pH was increased from 7.5 to 10.05 to effect elution of the adsorbed lactase.

Affinity chromatography was used by Katz and co-workers at New York University to isolate lysozyme from human fluids. A derivative of chitin—a compound obtained from squid backbone—was used as the ligand. The insoluble chitin derivative was packed in a column and a homogenate from human tissues was passed through it. Lysozyme bonded specifically to the chitin. Other proteins flowed through the col-

---

[1]An angstrom is one ten-billionth of a meter.

umn unhindered and traces of these were washed out with distilled water. The lysozyme complex was dissociated by diluted acetic acid. Average yields of 99.5% lysozyme were obtained. The enzyme was 3.2 times as active as the purest commercially available lysozyme. The enzyme can be isolated in 2 or 3 hours by use of this technique. Previous methods took up to 3 months and the product was less pure.

ENZYME INHIBITORS IN WAR AND AGRICULTURE

Nerve fibers carry electrical messages from the brain to cause muscles to contract. The end of the nerve fiber is separated from the muscle it stimulates by a gap of about 10^{-8} meter (one hundred-millionth of a meter). When the electrical message reaches the end of the nerve cell, it causes the production of a chemical known as acetylcholine. This chemical crosses the gap and depolarizes the muscle cell membrane much as does an electrical shock. The muscle contracts in response to this stimulus. At the muscle surface, acetylcholine is rapidly hydrolyzed by the enzyme acetylcholinesterase. The decomposition of acetylcholine is illustrated in Equation 15–3.

Equation 15–3

$$H_3C^+\ NCH_2CH_2OCCH_3 \text{ and } H_2O \xrightarrow[\text{Enzyme}]{\text{Acetyl-cholinesterase}}$$

$$\begin{matrix} H_3C & CH_3 \\ Cl^- \end{matrix}$$

Acetylcholine chloride and Water

$$H_3C^+\ NCH_2CH_2OH \text{ and } CH_3COOH$$

$$\begin{matrix} H_3C & CH_3 \\ Cl^- \end{matrix}$$

Choline chloride
and Acetic acid

This stops the stimulation and allows the muscle to relax. Before another contraction can occur, another nerve stimulation and acetylcholine release is required. But a normal muscle-nerve system can take part in hundreds of these processes—stimulation, contraction, and relaxation —every second.

If the acetylcholine is not decomposed, the muscle will remain in the contracted state and additional stimuli will release additional acetylcholine producing continuous contraction (tetanus). This process can take place at every nerve junction at every muscle, gland, and organ in the body and it can be brought about by anything which inhibits the enzyme, acetylcholinesterase.

Symptoms of such conditions include rapid twitching of muscles followed by paralysis, intense pupil contraction, excessive saliva production, and loss of coordination. The end result is death.

Organic esters of phosphoric acid (organophosphorus compounds) are potent inhibitors of acetylcholinesterase and this class includes the infamous "nerve gases," which were prepared but, fortunately, not used during World War II.

The nerve gases include the so-called "G" and "V" agents. The G agents are colorless liquids under normal conditions, but are very volatile. They are readily absorbable through the lungs, eyes, intestinal tract, and skin. Death can be caused in minutes. The V agents are less volatile, but may be aerosolized. They can produce casualties when absorbed through the skin in concentrations much lower than those required by G agents. They are lethal if inhaled.

As early as 1930, organophosphorus compounds were found to have insecticidal activity. But the nervous system of insects has not been studied as much as the mammalian system and it is not certain how these compounds kill insects. There is evidence that it is not by acetylcholinesterase inhibition. Some evidence indicates that certain insects remain healthy with all of their acetylcholinesterase inhibited and other insects, killed by pesticides, have been found to have much of their acetylcholinesterase intact. It seems likely that the insecticidal activity is due to enzyme inhibition, but the vital enzyme in insects is probably not acetylcholinesterase.

Because of the possible military use of organophosphorus compounds and because of the potential pollution problems, extensive research has been conducted into methods of detecting low levels of these enzyme inhibitors. Much of this work has been done at the Edgewood Arsenal near Baltimore. Most approaches to detecting these agents are based on enzyme inhibition. A device using immobilized cholinesterase for the rapid detection of enzyme inhibitors in air or water has been described by Dr. Louis Goodson of the Midwest Research Institute who worked with Mr. John Young and Mr. Andrew Davis of Edgewood Arsenal in devising the detector. Some organophosphorus nerve gases and insecticides are shown in Figure 15–1.

Figure 15-1   Some organophosphorus nerve gases and insecticides.

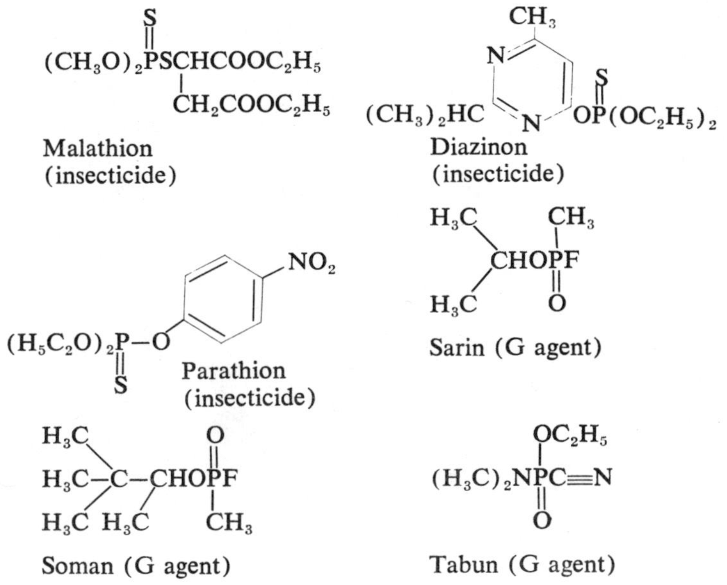

$(CH_3O)_2\overset{\overset{\displaystyle S}{\|}}{P}SCHCOOC_2H_5$
$\qquad\qquad\quad CH_2COOC_2H_5$

Malathion
(insecticide)

Diazinon
(insecticide)

$(H_5C_2O)_2\overset{\|}{\underset{\displaystyle S}{P}}-O$

Parathion
(insecticide)

Sarin (G agent)

Soman (G agent)

Tabun (G agent)

THERMAL STUDIES OF ENZYME REACTIONS

Nearly all chemical reactions either liberate or absorb heat. Enzymic reactions are no exception. Dr. Sam Pennington of East Carolina University has used an instrument called a microcalorimeter to measure the heat evolved in a number of enzyme-catalyzed reactions. Data obtained by this technique are useful for theoretical interpretation and it has practical application in analysis.

Dr. Pennington studied the acetylcholine-acetylcholinesterase system in the presence and absence of an organophosphorus-type inhibitor. As expected, there was considerably more heat evolved in the absence of the inhibitor. But, unexpectedly, the initial rate of the reaction in the presence of the inhibitor was faster than that of the noninhibited reaction. This was interpreted as meaning that the organophosphorus compound served as an alternate substrate for the enzyme, and was hydrolyzed, but remained bonded to the enzyme. A single calorimetric measurement gives data useful in measuring the rate of reaction, as well as in determining unknown concentrations.

Dr. Pennington has improved the microcalorimeter by using immobilized enzymes as part of the detector. This is intended to give good sensitivity while permitting more rapid measurements. The detector is, in effect, an enzyme thermistor.

This concludes our discussion of ways in which enzymes have been used by man. It has not been encyclopedic, but has been intended to serve as an introduction. We shall now turn to that most hazardous of all endeavors—predicting the future.

Chapter Sixteen

Speculation on the Future of Enzyme Technology

PREDICTING THE FUTURE is, indeed, a hazardous endeavor. Predicting the future development of a specific area of science is particularly difficult because of external influences—political, military, and economic —as well as scientific influences.

An important advance in another field may make today's impossible dream tomorrow's achievement. A classic example of this was provided by X-ray studies of protein structure. The studies were begun with the knowledge that they could not be completed in the lifetime of the original investigator. The calculations would require several lifetimes. But the computer was invented and these calculations could be completed in minutes.

But such unexpected events cannot be considered. We must base our anticipation of things to come on things that already are. The speculation in this chapter is therefore based on research—much of it long-range research—that is already going on, or is at least being discussed.

Immobilized enzymes have established themselves as useful tools in industry and analysis and have shown promise in medical applications. Immobilized whole cells have been used industrially. It seems likely that industrial uses of immobilized enzymes such as lactase, glucose isomerase, and glucoamylase will be important processes in the near future. This should encourage further industrial exploitation of these useful catalysts.

Research continues on cofactor regeneration. This problem will be solved, making available many more types of enzymes for industrial, analytical, and medical use.

Fermentation should expand, not only because of its likely role as a source of high-quality protein for foods and feeds, but also because it provides an excellent source of enzymes for many uses. Genetic manipulation can increase an enzyme content of a given organism until that enzyme comprises 20% of the organism's protein. Thermophilic (heat-loving) bacteria provide enzymes that are very stable to heat. These may be used at elevated temperatures giving more rapid conversions of substrates.

The production of organic solvents and other products, which were taken over by the petroleum industry, may once again be achieved by fermentation. The economic situation has changed considerably since fermentation was driven to find new worlds to conquer.

The conversion of waste materials to useful products by microorganisms or enzymes will become commercially important. The conversion of cellulose-containing materials (such as wastepaper) to glucose by cellulase is an example of such a process.

A combination of solar radiation and enzyme technology will be used to produce both food and energy in future years. The production of methane by microorganisms will be increased to provide much of the gas consumed as fuel. Hydrogen gas may also be produced by a process known as biophotolysis—splitting the water molecule by a combination of light and enzyme action. This is essentially an adaptation of photosynthesis to man's design.

Biological methods for "fixing nitrogen"—taking it out of the air and converting it to a usable form—will be developed. This will not only provide fertilizers, but will also be useful in production of food and fuel.

Enzymic methods of analysis should replace some of the older methods. The use of enzymes makes the assay method more specific and, in many cases, more rapid and reliable. Devices, such as enzyme electrodes, will gain in popularity, just as specific-ion electrodes have.

The ever-increasing number of carefully controlled studies will eventually convince physicians that modern, highly purified enzyme preparations are safe, reliable, and effective against many forms of malady. Relatively inexpensive enzymes may replace more costly medicaments.

Asparagine may be just one of many nutrients that are essential to some cancer cells, though not to normal cells. Glutamine has already been added to the group. Many more of these may be found and they may be destroyed by a specific enzyme, as we have seen in the asparaginase example.

The use of proteolytic enzymes, such as the Wobe-Mugos preparation, will end much of the suffering that accompanies neoplastic diseases (cancer). Many patients will be cured and many more will die a less agonizing death. Metastases due to surgical or X-ray treatments will be reduced by enzyme adjunct therapy and the rate of cancer cures will increase dramatically. It will no longer be necessary to resort to potentially lethal drugs, such as methotrexate.

Enzyme replacement therapy by use of organ transplants will probably become a more acceptable technique and will be helpful to more patients. Methods for temporarily opening the blood-brain barrier will be developed making enzyme therapy available for enzyme deficiency diseases involving the central nervous system. Genetic engineering—already being developed in microorganisms—will, in the distant future, provide the ultimate answer to curing such diseases.

Solid-phase synthesis methods are being constantly improved. Automated synthesis of pure proteins in large quantities should be possible. The technique will be useful not only for synthesizing enzymes, but other complex molecules as well.

Enzymes will become increasingly familiar to the synthetic organic chemist. They will be useful in the synthesis of natural products and also in simpler conversions. Their usefulness under mild conditions will help overcome the chemist's fear of the unknown.

CONCLUSION

Some of these speculations will prove correct; others will not. Discoveries in other fields will expand the possibilities for enzyme uses, just as enzymes will expand other fields. Whatever the direction taken by enzyme technology, it is here to stay and you now have a basic understanding of the field.

Bibliography

"If I have been able to see farther than others, it was because I stood on the shoulders of giants." These words of Sir Isaac Newton tell much of how science advances. In this book, we have not always been concerned with identifying the "giants" who first suggested a given enzyme application, nor those who conducted the theoretical investigations, which precede most technological advances. We have been more concerned with mentioning specific features of a given application, so that the investigators named were not, in some cases, the originators of a given application. For the benefit of those interested in learning of the pioneering giants, as well as for those who wish to know more about enzymes and their uses, a bibliography is appended.

In general, the organization of the bibliography follows that of the book. The books by Locke and by Moss, listed in the general section, as well as the one by Asimov listed under "Photosynthesis," require no particular technical background. The rest of the books and all of the review articles are more technical and were written for those having a scientific background.

BOOKS

General

Dunlap, R. B., ed. 1974. Immobilized biochemicals and affinity chromatography. *Advances in experimental medicine and biology.* No. 42. New York: Plenum Press.

Locke, D. M. 1969. *Enzymes—The agents of life.* New York: Crown Pub.
Meltzer, Y. L. 1973. *Encyclopedia of enzyme technology.* Flushing, N.Y.: Future Stochastic Dynamics, Inc.
Moss, D. W. 1968. *Enzymes.* London and Edinburgh: Oliver & Boyd.
Tribe, M. A. 1976. *Enzymes.* New York: Cambridge University Press.
Tribe, M. A., Eraut, M. R., and Snook, R. K. 1976. *Metabolism and mito-chondria.* New York: Cambridge University Press.
Walter, C. 1976. *Enzyme reaction and enzyme systems.* New York: Dekker.
Williams, A. 1969. *Introduction to the chemistry of enzyme action.* New York: McGraw-Hill.
Wiseman, A. 1971. *Enzymes: Their nature and role.* London: Hutchinson Educational.
Zuber, H., ed. 1976. *Enzymes and proteins from thermophilic microorgan-isms.* Basel: Birkhauser.

Analytical Uses Including Clinical Analysis

Blume, P., and Freier, E. F., eds. 1974. *Enzymology in the practice of laboratory medicine.* New York: Academic Press.
Coodley, E. F., ed. 1970. *Diagnostic enzymology.* Philadelphia: Lea & Febiger.
Guilbault, G. G. 1970. *Enzymatic methods of analysis.* Elmsford, N.Y.: Pergamon Press.
Wilkinson, J. H. 1976. *Diagnostic enzymology.* London: Edward Arnold.

Bioluminescence

Klein, H. A. 1965. *Bioluminescence.* Philadelphia: Lippincott.

Enzyme Synthesis

Stewart, J. M. 1969. *Solid phase peptide synthesis.* San Francisco: W. H. Freeman.

Food and Dairy Uses

Olson, A. C., and Cooney, C. L. 1974. *Immobilized enzymes in food and microbial processes.* New York: Plenum Press.
Ory, R. L., and St. Angelo, A. J., eds. 1977. *Enzymes in food and beverage processing.* Washington, D.C.: Chemical Society Press.
Reed, G. 1975. *Enzymes in food processing.* 2d ed. New York: Academic Press.
Whitaker, J. R. 1972. *Principles of enzymology for the food sciences.* New York: Dekker.

Immobilized Enzymes

Gutcho, S. 1974. *Immobilized enzymes.* Park Ridge, N.J.: Noyes Data Corp.
Mosbach, K., ed. 1976. *Methods in enzymology, 46: Immobilized enzymes.* New York: Academic Press.

Salmona, M., Soronio, C., and Garattini, S., eds. 1974. *Insolubilized enzymes.* New York: Raven Press.
Weetall, H. H., ed. 1975. *Immobilized enzymes, antigens, antibodies and peptides: Preparation and characterization.* New York: Dekker.
Weetall, H. H., and Suzuki, S., eds. 1975. *Immobilized enzyme technology.* New York: Plenum Press.
Zaborsky, O. 1973. *Immobilized enzymes.* Cleveland: CRC Press.

Industrial Uses

Bohak, Z., and Sharon, N., eds. 1977. *Biotechnological applications of proteins and enzymes.* New York: Academic Press.
Gaden, E., ed. 1976. *Enzymatic conversion of cellulosic materials.* New York: Interscience.
Johnson, J. C. 1977. *Industrial enzymes recent advances.* Park Ridge, N.J.: Noyes Data Corp.
Messing, R. A., ed. 1975. *Immobilized enzymes for industrial reactors.* New York: Academic Press.

Medical Uses

Bondy, P. K., and Rosenberg, L. E., eds. 1974. *Genetics and metabolism.* Duncan's diseases of metabolism, 7th ed., vol. 2. Philadelphia: W. B. Saunders Co.
Chang, T. M. S. 1972. *Artificial cells.* Springfield, Ill.: Thomas.
Chang, T. M. S., ed. 1977. *Biomedical applications of immobilized enzymes and proteins.* New York: Plenum Press.
deReuck, A. V. S., and Cameron, M. P., eds. 1963. *Lysosomes.* Ciba Foundation Symposium. Boston: Little, Brown.
Garrod, A. E. 1923. *Inborn errors of metabolism.* 2d ed. New York: Oxford University Press.
Innerfield, I. 1960. *Enzymes in clinical medicine.* New York: McGraw-Hill.
Lauwers, A., Sharpe, S., Cooreman, W., and Ruyssen, R. 1974. *Pharmaceutical enzymes.* Ghent, Belgium: E. Story—Scientia P.V.B.A.
Mammen, E. F., Anderson, G. F., and Barnhardt, M. I., eds. 1971. *Thrombolytic therapy.* Stuttgart: F. K. Schattauer Verlag.
Mandl, I., ed. 1972. *Collagenase.* New York: Gordon & Breach.
Martin, G. J. 1958. *Clinical enzymology.* Boston: Little, Brown.
Seegers, W. H. 1967. *Blood clotting enzymology.* New York: Academic Press.
Stanbury, J. B., Wyngaarden, J. B., and Frederickson, D. S., eds. 1972. *The metabolic basis of inherited disease.* 3d ed. New York: McGraw-Hill.
Wolf, M., and Ransberger, K. 1972. *Enzyme therapy.* New York: Vantage Press.

Photosynthesis

Asimov, I. 1968. *Photosynthesis.* New York: Basic Books.
Rabinowitch, E., and Govindjee, X. 1969. *Photosynthesis.* New York: John Wiley.

Tribe, M. A. 1975. *Photosynthesis.* New York: Cambridge University Press.

REVIEW ARTICLES

Brown, H. D., and Hasselberger, F. X. 1971. Matrix supported enzymes. In *Chemistry of the cell interface,* part B, H. D. Brown, ed., pp. 185–258. New York: Academic Press.

Denkewalter, R. G., and Hirschmann, R. 1969. The synthesis of an enzyme. *American Scientist* 57(4):389–409.

Gregoriadis, G. 1976. The carrier potential of liposomes in biology and medicine. *New England Journal of Medicine* 295:704–10; 765–70.

Gryszkiewicz, J. 1971. Insoluble enzymes. *Folia Biologica* 19(1):119–50.

Katchalski, E., Silman, I., and Goldman, R. 1971. Effect of the micro-environment on the mode of action of immobilized enzymes. In *Advances in enzymology,* vol. 34, F. F. Nord, ed. New York: Interscience Pub.

Marglin, A., and Merrifield, R. B. 1970. Chemical synthesis of peptides and proteins. *Annual Review of Biochemistry* 39:841–66.

McLaren, A. D., and Packer, L. 1970. Some aspects of enzyme reactions in heterogeneous systems. In *Advances in enzymology,* vol. 33, F. F. Nord, ed. New York: Interscience Pub.

Orth, H. D., and Brummer, W. 1972. Carrier-bound biologically active substances and their applications. *Angewandte Chemie* (int. ed.) 11: 249–346.

Raab, W. P. 1972. Diagnostic value of urinary enzyme determinations. *Clinical Chemistry* 18:5–25.

Silman, I. H., and Katchalski, E. 1966. Water-insoluble derivatives of enzymes, antigens and antibodies. *Annual Review of Biochemistry* 35 (part 2):873–908.

Suckling, C. J., and Suckling, K. E. 1974. Enzymes in organic synthesis. *Chemical Society Reviews* 3(4):387–406.

Weetall, H. H., and Messing, R. A. 1971. Insolubilized enzymes on inorganic materials. In *Chemistry of bio-surfaces,* vol. 2, M. L. Hair, ed., pp. 563–95. New York: Dekker.

Index